12

Short-Time Compensation

Pergamon Titles of Related Interest

Guzzo/Bondy/Work in America Institute A Guide to Worker Productivity
   Experiments in the United States 1976–81
Klein Prices, Wages and Business Cycles
Lundstedt/Colglazier Managing Innovation: The Social Dimensions of
   Creativity, Invention and Technology
O'Brien/Dickinson/Rosow Industrial Behavior Modification: A Management
   Handbook
Work in America Institute Productivity Through Work Innovations: A Work
   in America Institute Policy Study
Zager/Rosow The Innovative Organization: Productivity Programs in
   Action

Related Journals

Work in America Institute Studies in Productivity
World of Work Report

Short-Time Compensation

A Formula for Work Sharing

Edited by
Ramelle MaCoy
Collective Bargaining Associates
Martin J. Morand
Indiana University of Pennsylvania

Pergamon Press
New York • Oxford • Toronto • Sydney • Frankfurt • Paris

Pergamon Press Offices:

U.S.A.	Pergamon Press Inc., Maxwell House, Fairview Park, Elmsford, New York 10523, U.S.A.
U.K.	Pergamon Press Ltd., Headington Hill Hall, Oxford OX3 0BW, England
CANADA	Pergamon Press Canada Ltd., Suite 104, 150 Consumers Road, Willowdale, Ontario M2J 1P9, Canada
AUSTRALIA	Pergamon Press (Aust.) Pty. Ltd., P.O. Box 544, Potts Point, NSW 2011, Australia
FRANCE	Pergamon Press SARL, 24 rue des Ecoles, 75240 Paris, Cedex 05, France
FEDERAL REPUBLIC OF GERMANY	Pergamon Press GmbH, Hammerweg 6, D-6242 Kronberg-Taunus, Federal Republic of Germany

Library of Congress Cataloging in Publication Data

MaCoy, Ramelle.
    Short-time compensation.

    Includes bibliographical references and index.
    1. Work sharing.   2. Insurance, Unemployment.
I. Morand, Martin J. II. Title.
HD5110.5.M32 1984   331.25′72   83-13265
ISBN 0-08-030148-7

Printed in the United States of America

Pergamon Press/Work in America Institute Series

Work in America Institute, Inc., a nonprofit, nonpartisan organization, was founded in 1975 to advance productivity and quality of working life. Through a series of policy studies, education and training programs, an extensive information resource, and a broad range of publications, the Institute has focused on the greater effectiveness of organizations through the improved use of human resources. Because of its close working relationships with unions, business, and government, the Institute is sensitive to the views and perspectives of all parties and is recognized as an objective source of information on issues of common interest.

The Pergamon Press/Work in America Institute Series is designed to explore the role of human resources in improving productivity in the workplace today and to identify the trends that will shape the workplace of the future. It will focus primarily on the issues of:

- *Quality of working life* through work innovations that encourage employee participation in decision making and offer recognition to employees for their contributions.
- *Productivity* through the more effective management of human resources.
- *Interaction between people and technology* to achieve a more satisfactory transition to the workplace of the future.
- *Labor-management cooperation* to solve mutual problems in the workplace, to achieve improved quality of product, and to make organizations work better.
- *National labor-force policy* as it relates to productivity and the quality of working life.

Contents

Foreword

During the 1970s many new developments caused widespread change in the American workplace. The influx of women into the paid work force, a dramatic increase in single-parent and dual-career families, a growing preference of workers to trade money for time off, continued emphasis on equal employment opportunity for minorities and women, a desire for greater employee participation in the workplace, and the crisis caused by declining American productivity and increased trade competition from abroad were factors that encouraged experimentation with new ways of scheduling work.

An increasing number of organizations around the world have introduced flexible work scheduling in both offices and factories for workers who have traditionally worked rigid schedules. There has been widespread acceptance of alternative work schedules in the private sector in the United States since flexitime was first introduced in 1967 at a German aerospace firm to alleviate transportation delays that prevented employees from getting to work on time.

As chairwoman of a House Post Office and Civil Service Subcommittee, I held hearings on legislation that was enacted in 1978 to permit federal agencies to experiment with flexible and compressed work schedules and to increase permanent part-time career employment in government. The response to these programs has been overwhelmingly positive. The improved employee morale, higher productivity, and longer hours of service for the public have benefited both employee and employer. It was because of my legislation involving flexitime, the four-day week, and permanent part-time employment that I became interested in short-time compensation (STC), a partial payment of the weekly unemployment benefits to compensate workers whose hours have been reduced by a work-sharing plan due to recession. I believed that short-time compensation offered an acceptable alternative to layoffs.

In June 1980, I introduced legislation to promote STC, because this is an *employment* program that utilizes the unemployment compensation system to keep people attached to their jobs during lean times. It is not appropriate for all jobs, all workers, or all businesses at all times, but where it is appropriate, an STC program can be an enormous help to a firm and its employees. STC gives us another tool for dealing with recessions and prevents the dislocation of workers.

When I first introduced my STC legislation in 1980, we were in yet another economic turndown, with thousands of workers laid off in the automobile, rubber, steel, and construction industries. There has been a growing concern that recessions do not affect all segments of society equally, that the brunt is borne by those employees who are laid off— usually the newly hired, least skilled, minority and female workers. The physical and psychological consequences of long-term layoffs are devastating—loss of income, fringe benefits, job status, self-esteem, and an increase in alcoholism, health problems, child abuse, and the deterioration of family life. Layoffs also depress local economies by reducing not only the purchasing power of laid-off workers but also the consumer spending of employed workers who fear potential layoffs.

Those laid-off workers are not the only ones hurt by recession. Revised production schedules and bumping procedures often mean that those workers who are not laid off end up in lower-paying jobs that they find unsatisfactory. The companies with layoffs find significant reductions in productivity resulting from the high cost of reorganization, changed production schedules, and the hiring and training of new workers.

The taxpayer is also hurt by layoffs. The Congressional Budget Office estimates that for each additional percentage of unemployment, federal entitlement programs cost $5 to $7 billion more and tax revenues decline by $20 to $22 billion. Moreover, recessions have contributed to the current Social Security financing crisis, since unemployment reduces the amount of payroll taxes paid into the Social Security trust funds.

I introduced my bill for two reasons: to encourage states to experiment with STC and to make it a priority issue for the Department of Labor. Ideally, the federal government should encourage the states to experiment with new ways to solve problems. The Feds should also evaluate state experiments for future application in national economic policymaking.

My legislation was drafted to protect workers by limiting STC benefits for a worker to 26 weeks during the year, by requiring the approval of the union representative before an STC plan could be initiated, and by continuing fringe benefits for workers on STC. The guidelines in the bill, which were based on the experience of the California program, were recommended to the states, although they were not made mandatory for them. Since this is a new program, states should have leeway to experiment with STC. The federal evaluation should examine the differences in state laws and their effects to determine what works best.

The bill was designed to protect the solvency of the unemployment insurance (UI) trust fund by requiring that employers certify that the use of STC is in lieu of an equivalent number of hours of full-time layoffs, and that STC be financed according to state law. In California, Arizona, and Oregon, negative-balance employers, who pay higher UI taxes per employee because of frequent use of the UI system, are charged a surtax on their use of STC. In California, the surtax provision limited the number of negative-balance firms using the STC program. It is important that the federal evaluation carefully assess whether financing mechanisms are effective in protecting the solvency of the UI trust funds by providing for adequate financing of the STC benefits by employers who use work sharing. At the same time, we do not want to make the use of STC so expensive for employers affected by cyclical downturns that they are discouraged from using it. If the cost of full-time layoffs is the same as the use of STC in its effect on the trust funds, there should be no built-in disincentives to employers who would otherwise consider work sharing instead of layoffs.

STC is not a substitute for prudent and efficient resource management to minimize the necessity of layoffs. Nor is it a cure for unemployment, although it can reduce the number of layoffs necessary when the economy is not as healthy as it should be. STC is not a substitute for other jobs and training programs or for the regular unemployment insurance system. However, the availability of STC does give us another weapon in our arsenal with which to fight unemployment.

An STC program will only succeed with support from the firm, its workers, and their union. One of the side benefits of the STC program is that it has promoted cooperation between management and unions in finding alternatives to layoffs. From experience in the states, STC offers us a model for improved labor relations, since it focuses on co-operation in solving the serious problems created by cyclical downturns.

Patricia Schroeder
Member of Congress
First Congressional District, Colorado

Preface

This book is a timely and informative analysis of short-time compensation (STC), a new option for combating joblessness in an industrial society. The combination of a shorter workweek with partial unemployment insurance benefits for lost work time constitutes a social invention with important economic, social, and political benefits to employers, employees, the community, and the nation.

Work in America Institute is pleased to sponsor this important book in order to make available to a much wider audience the tools for establishing short-time compensation as an attractive alternative to layoff. The Institute's interest in STC has been manifested through several of its own activities. The policy study report, *New Work Schedules for a Changing Society*, released in 1981, proposed this alternative as one of several new ways of managing working time in a manner more compatible with the needs of the organization, the life-styles of employees, and the state of the economy. At present, the Institute is engaged in another national policy study, "Employment Security in a Free Economy," which will outline a diverse set of options preferable to layoff, including short-time compensation. The report will be released in 1984.

Short-Time Compensation: A Formula for Work Sharing, edited by Ramelle MaCoy and Martin Morand, reviews short-time compensation in considerable depth. It discusses the pioneering programs of the Federal Republic of Germany; the STC programs of three innovative states: California, Arizona, and Oregon; and Canada's new and thriving form of short-time compensation. More than that, the book views STC from the perspective of both management and labor and analyzes its effect on labor-management cooperation and affirmative action. Other policy issues are discussed in the final chapter. All the evidence supports the fact that these programs are cost effective, popular with organizations and employees, and deserving of wider attention and application in the United States.

Rep. Patricia Schroeder, who has contributed an interesting foreword to this book, spearheaded the effort in the federal government to advance this option to most states through legislation that encourages them to undertake their own STC programs. Her legislative initiative, Public Law 970248, required the Secretary of Labor to develop model legislation that may be used by the states to enact short-time compensation programs and also to provide them with technical assistance in developing and implementing such programs. It was signed into law in September 1982. Recent legislative actions in the states include the passage of an STC law in Florida and Washington and partial passage in Illinois of an STC law (it passed the Senate and is pending in the House). In addition, the New York State Legislature has had an STC law under consideration for several years.

Although this exciting concept is still in its infancy in North America, it has benefited about 1.2 million people in the Federal Republic of Germany during the current recession. It has succeeded to such a degree that it merits the positive consideration of state governments and business and union leaders everywhere. We believe that MaCoy and Morand's study of short-time compensation will be instrumental in bringing this concept to the attention of these policymakers as well as the general public.

We have been assisted in this project by a grant from The German Marshall Fund of the United States. We appreciate their support and note that the views expressed herein are those of the writers and do not necessarily represent the views of the fund. We also thank the distinguished contributors to this book, who have offered us insights on STC based on their familiarity with specific programs and expertise in their specialized fields.

Jerome M. Rosow
President
Work in America Institute

Acknowledgments

We are indebted to the large number of persons without whose contributions this book would not have been completed.

Little need be said regarding the authors of specific chapters. Their work speaks for itself and demonstrates both their knowledge and insight and their willingness to shed light on a topic they deem of great social interest and importance.

To most of our many colleagues and assistants a brief thank you is all that space will permit, but a special word is in order in at least some few cases.

In the beginning there was a concept paper "Encouraging Alternatives to Seniority-Based Layoffs through Modifications of Unemployment Insurance" co-authored by Dr. Donald S. McPherson, chairperson of the Department of Industrial and Labor Relations of Indiana University of Pennsylvania (IUP). McPherson's inspiration and assistance continued with the research reported at the First National Conference on Work Sharing Unemployment Insurance, San Francisco, May 1981, as "Union Leader Responses to California's Work Sharing Unemployment Insurance System."

This report and further research were made possible through the generous support and encouragement of IUP. The initial suggestion to conduct this research was triggered by a call from an old friend from the early sixties civil rights movement, Robert Rosenberg, who had provided the policy analysis which led to passage of the first American short-time compensation legislation. Fred Best, who headed the research team conducting the interim evaluation for the California Employment Development Department, was most generous with his time and knowledge of the topic. The department's then director, Douglas Patino, provided moral and logistical support for our research and continues in his new position as director of the Department of Economic Security in Arizona, to play a leading role in research on, and expansion of, STC nationally. Senator Bill Greene, who shepherded the original STC bill through the California legislature, continues to give legislative oversight to the issue and is supportive of ongoing research efforts. Greene and Patino combined legislative and executive branch resources to create the First National Conference, whose success inspired our decision to undertake this volume.

A few of the other Californians whose help was critical and acknowledged thus briefly are: Joan Bissel, Dotty Block, Barbara Chrispin, Gera Curry, Masako Dolan, Hellan Dowden, Audrey Gardner, Carol Holmes, Frank Kessling, Richard Quaresma, Mike Rice, Werner Schink, Norman Skonovd, Stephen Story, Dean Tibbs, and Casey Young.

At IUP there were faculty colleagues, graduate assistants, and other students whose ideas and assistance are valued and appreciated. They are: Marilyn Baker, Rosemary Bennett, Patrick Boyle, James Dyal, Michael Haberberger, Nicholas Karatjas, Sharon Lenart, Jack LeVier, Irwin Marcus, Susan Morand, Tom Nowak, George Pazul, Linda Roberts, Kay Snyder, and Robert Spear.

The persons who typed drafts provided far more than what is commonly called "clerical assistance." They corrected spelling and grammar, read what they typed, and asked questions on the assumption that if they couldn't make sense out of something, neither could the reader. They are: Audrey Cornrich, Cherie Haubert, Theresa Logan, Christina Meyers, Mary Beth Nobers, Dawn Oberholtzer, and Kathleen Walters.

Finally and essentially, Work in America Institute, as sponsor, provided the support that made this volume possible. It was the encouragement and editorial skill of Beatrice Walfish of the Institute, however, that turned this collection of chapters into an integrated book.

Our thanks to all.

R. MaCoy
M. Morand

I
SHORT-TIME COMPENSATION: AN ALTERNATIVE TO LAYOFF

1.
Work Sharing and Short-Time Compensation: An Overview

Ramelle MaCoy
Martin J. Morand

In the late twentieth century, full employment as we have understood this term remains an elusive goal. In the world of work, the availability of sophisticated technology and increased labor costs are encouraging employers to replace labor with automated and computerized machinery. These circumstances, in combination with recurring recessions, are producing progressively greater numbers of unemployed. At the same time, increased fixed living costs are, for the individual, exacerbating the traumatic impacts of even temporary periods of unemployment.

This book deals with a new (for America) way to work which some firms and workers are experiencing and which may give direction to the way in which society adapts to the radically changed economic circumstances of this period. It is called short-time compensation (STC)—work sharing supplemented by shares of regular unemployment insurance (UI) benefits prorated in proportion to the reduction of work time.

This system of encouraging employment instead of subsidizing unemployment was initiated in Germany over 50 years ago and, with variations, is in place in most of western Europe and Japan. Three American states—California, Arizona, and Oregon—have recently amended their UI laws to permit STC benefits for reduced workweeks, and Canada has adopted a similar program.

The concept of STC is best illustrated by a simple example. Suppose, because of reduced demand, the employer of 100 workers is planning to lay off 20 of them. Instead, however, he reduces the workweek of all 100 workers by 20 percent, and the same UI benefits that would have been paid to the 20 laid-off workers are divided among the 100. Even though there would be a slight reduction in earnings for all 100 workers, none would suffer the traumatic hardship of total unemployment; the employer would maintain a balanced, skilled, and productive work force; and there would be no significant additional drain on the UI fund.

Sharing the available work can take many forms, depending on the nature of the organization and its productive processes. The most common is reduction of the workweek, usually from five days to four. Other options include reducing the hours of work per day, shutting down the entire plant for a week or more, and alternating or rotating layoffs. In the latter arrangement, part of the work force is laid off for a period, after which it returns to work, and another group is laid off for a similar period.[1]

California, which adopted STC in 1978, was the only American state to experiment with the program until Arizona in 1981, Oregon in 1982, and Washington and Florida in 1983 adopted similar legislation. The titles given the program vary among nations and among states (in California it is called Work Sharing Unemployment Insurance or WSUI), but we shall use *short-time compensation* (STC), the title adopted by Congress, as a generic term for all such programs.

Benefits of Short-Time Compensation

Even before Arizona and Oregon joined California in experimenting with STC, the concept was carefully studied by labor-market experts and widely discussed in government, labor, and management circles. A thoughtful and concise analysis of the impact of unemployment on society, and the potential benefits of short-time compensation, was developed in 1981 by Work in America Institute in its national policy study *New Work Schedules for a Changing Society.*[2]

Each time large-scale layoffs occur, says the Work in America Institute study, society undergoes severe shocks:

- Laid-off workers suffer loss of wages, benefits, security, morale, skills, work habits, and physical and psychic health. Many remaining workers are "bumped" into lower-paid jobs.
- Family ties are strained—offset, in part, by the fact that more families now have multiple wage earners.
- Payments on mortgages and cars are not met.
- Layoffs hurt the employer, too. Productivity is lost due to the reworking of production schedules and the "bumping" of workers.
- Unemployment insurance premiums increase.
- Violence and crimes increase.
- Each 1 percent increase in unemployment is reflected in a 2 percent increase in the mortality rate within five years.
- Communities take on heavier costs for health, welfare, police, and other services, while revenues diminish.

- Each 1 percent increase of unemployment costs the federal government $20 billion of benefit payments and lost revenues.
- Unemployment insurance and Social Security funds are drained.
- The federal budget moves further into deficit, and the government is pressed to take measures that permanently distort the economy.

In most cases, especially where employees are unionized, the order of layoff and recall is determined by length of service or seniority under the principle of last in, first out. Women, youth, and members of minority groups make up a disproportionate number of the victims. Many cannot afford to wait to be recalled, so they seek jobs elsewhere, begin again with zero seniority, and remain permanently vulnerable.[3]

Work sharing with short-time compensation can alleviate many of these pressures, says the Institute, but the rigidity of the unemployment system remains the impediment. It maintains that if this impediment could be removed and if large numbers of employers and employees were to adopt work sharing in place of layoffs, these advantages would flow:

For employers
1. Maintenance of productivity because of higher morale and preservation of employee skills.
2. Retention of skilled workers.
3. Reduction or elimination of the large costs associated with layoffs, particularly where "bumping" occurs, for example, distorted production scheduling, delayed start-ups when recession ends, retraining of bumped employees.
4. Greater flexibility in deploying human resources to keep operations going.
5. Savings in employer costs associated with severance pay, early-retirement incentives, and other layoff schemes requiring substantial financing.
6. Avoidance of postrecession costs of hiring and training new workers to replace those who found other jobs during layoff.
7. Reinforcing group loyalties and strengthening employee loyalty to the firm.

For workers
1. Continued job attachment for workers who would otherwise have been laid off.
2. Continued fringe-benefit protection for employees and their families.
3. Retention of more minority and women workers, thus preserving the aims and achievements of affirmative action.
4. Security for older workers, who cost more, are often among the first fired in selective layoffs, and are discriminated against when they seek new jobs.
5. Opportunity for workers with high seniority to trade work for increased leisure, with only a small decrease in take-home pay, thus providing a "taste" of retirement without fear of unemployment.
6. More effective preservation of the family income of two-paycheck families than if one member continues to work full time and the other is on unemployment insurance, particularly if the wife is a new entry into the labor

force. (Women's traditional jobs are generally low paying, and eligibility rules for unemployment insurance require workers to meet a combination of dollar amount and time in covered occupations. For some new entrants, these benefits may be low or nonexistent.)

For unions
1. Preservation of union membership and members' ability and willingness to pay dues.
2. Greater ability to take into account diverse interests of membership and fairly represent all employees in the bargaining unit.
3. Improved long-run prospects for the union. Layoffs generally pose problems for unions because some laid-off workers do not return, new employees have to be organized, and returning workers may have less enthusiasm for the union after extended unemployment.
4. Less polarization between groups represented by the union.
5. Increased support from new workers who would otherwise be laid off.
6. Greater bargaining flexibility when an employer suffers a downturn.

For society
1. Protection of affirmative action and equal employment opportunity advances.
2. Less need for public-service jobs.
3. Less need for public assistance for the unemployed.
4. No increase in net unemployment insurance costs.
5. Less disruption of the society as a whole.[4]

An addendum: Whereas in this 1981 listing, "continued fringe benefits" are listed as an advantage "for workers," by 1983 awareness of the problems generated when health insurance is dependent on continued employment had transformed this into a societal issue. Health care providers were pressed to continue providing services to laid-off workers and their families. Health insurance carriers were offering cut-rate packages to their former customers who had been dropped from group plans. Tax dollars were funding increased numbers of Medicaid beneficiaries. Congress was considering emergency health care coverage for the unemployed. State legislatures were considering proposals to require employers to maintain health benefits for laid-off workers. New approaches to continuing postlayoff coverage including tripartite-funded insurance programs with worker, employer, and state contributions were being negotiated.

The assertion by one union leader that "the sicker you get, the poorer you get—the poorer you get, the sicker you get,"[5] along with the recognition that one uninsured illness could precipitate a family into a lifetime of indebtedness, suggested a new reason for supporting STC. When an employer adopts STC he is preserving the health insurance of more workers and, as *Forbes* noted, ". . . the state likes that because tax dollars probably would catch those health bills otherwise."[6]

Definitions

Work sharing, STC, job sharing, flexitime, and part-time work are all somewhat related, and it is important to distinguish among them.[7] In this book we shall attach the following meanings to these terms:

Flexitime is the adjustment of normal weekly hours to suit the convenience of an individual worker within guidelines set by the organization. A flexitime schedule may start or end before or after the normal hours of work; a single shift may be divided into two or more segments; and the hours per day may be increased or decreased in order to alter the number of days worked per week.

Part-time work is work permanently scheduled for fewer than the normal weekly hours. The Bureau of Labor Statistics defines part-time work as less than 35 hours per week, but it is more typically 20 hours or less.

Job sharing is the assumption of the responsibility for a single job by two workers who may divide the hours and days of the normal workweek in a variety of ways.

Work sharing is the temporary reduction in the working hours of a group of workers. A work-sharing program distributes a limited amount of work among the members of the group so as to avert the layoffs of a portion of the group. It may be achieved in a variety of ways, including a reduction in hours per day or days per week, or by the rotation of weeks worked and weeks off. The effect is to reduce the hours of the normal workweek. Work sharing was prevalent in the United States during the Great Depression, but those whose hours were cut were not compensated in any way for the time lost. After the introduction of unemployment insurance, this form of work sharing became obsolete.

Work sharing as used here does *not* refer to early retirement, worker sabbaticals, or extended holiday or vacation periods, even though any or all of these may have the similar intent and effect of preserving jobs through reductions in hours.

Short-time compensation (STC) is the payment of a pro-rata share of the regular weekly UI benefit to workers who experience a reduction in normal hours because of a work-sharing plan. STC is part of the regular UI system and is similar to, and a logical extension of, the more familiar "partial" UI benefit paid to workers who do a limited amount of work while unemployed. STC benefits are paid only to workers who participate in an approved work-sharing plan.

Short-time compensation is a labor-market policy intended to *preserve* employment through the avoidance of layoffs, at least in the short term. However, to the extent that the use of STC increases productivity through improvements in worker morale and the elimination of work disruptions caused by layoff—and there is some evidence that it does—some unemployment would logically be induced.

Any unemployment, thus induced, may ultimately be offset should STC prove to be a bridge toward a permanent reduction in the normal hours of work. If the increased efficiency of the shorter hours associated with STC proves to be more than a short-term Hawthorne effect, this alone would provide reason and support for a reduction in the forty-hour week. Much of the literature on fleximite has emphasized the economic benefits of the four-day week to workers, employers, and a society increasingly concerned about energy costs. Workers and unions experiencing the leisure-time benefits of STC can be expected to provide political support for proposals to shorten the normal workweek.

There is a difference between STC as a job preservation measure and a reduction in the regular hours of work as a means of job creation. The distinction is emphasized in the International Labour Office study of STC: "[STC] is completely different from reducing hours of work . . . with a view to *creating* employment. [STC] aims at preventing layoffs. . . ."[8]

Policies to stimulate growth and create employment are obviously an urgent priority in light of U.S. economic trends of the last decade: economic growth that averages under 2 percent, a decline in real wages, and unemployment now in the double-digit range. Since STC does not create employment, this has led some critics to demean it as a "Band-Aid." However, its cost is so minimal and its promises, by comparison, so great for workers, employers, and society that there seems little reason not to try it. The fact that its effects, unlike those of most employment policies, are immediate is further recommendation.

STC: 1927–1983

Since STC is a labor-market policy and part of the UI system, it tends to be identified as a worker benefit program and many people assume that it has been promoted by labor. The truth, both at home and abroad, is more interesting.

STC in Germany. It was apparently in Germany in 1927 that an unemployment insurance law was first modified to permit groups of workers operating under a planned reduced work schedule to collect pro-rata compensation benefits. Unemployment insurance itself had been instituted in Germany in the 1870s not in response to liberal or labor pressures but as a conscious and deliberate effort by the Iron Chancellor, Otto Von Bismarck, to woo the working class from its allegiance to the strong socialist movement of the day.

The entire European unemployment insurance system, including STC, has remained to this day suspect to German and other European socialist labor leaders who view it as a compromise at best, a temporary acknowledgment of the persistence of capitalism and the need to deal with its shortcomings, particularly unemployment.

The effectiveness of STC in "buying off" the working class was perhaps demonstrated in West Germany in 1983 when Chancellor Helmut Kohl's Christian Democratic Party, shortly before an election, extended STC benefits from 24 to 36 months. In reporting Kohl's subsequent election victory, the *New York Times*, on March 27, 1983, pointed out that for the first time in two decades, "in a monstrous shift, the blue collar belt rebuffed the Social Democrats. . . ." The *Times* pointed out that in Duisberg, the largest steel center in that nation, where 35,000 of 55,000 steelworkers are on short time, the extension of STC "guaranteed sixty-eight percent of their lost wages for another year." The conservative mayor of Duisberg, according to the *Times* story, asserted that the "gesture demonstrated a concern for steelworkers that paid off."

STC has always been introduced abroad as a labor-market tool to counter unemployment and sustain purchasing power through the alternative use of funds normally earmarked for laid-off workers. In the United States, however, it has a unique genesis and was originally advanced not out of concern about layoffs per se, but rather because of the disparate impact that *any* layoffs were presumed to have on recently hired women and minorities.

Seniority vs. Affirmative Action. The Civil Rights Act of 1964 was the greatest accomplishment of a liberal–labor–civil rights coalition, but ironically that act contained the seeds of destruction of the very coalition that lobbied it into existence. The then-president of the AFL-CIO, George Meany, played a central role in the passage of the act. It was George Meany who insisted that the provisions of the legislation go beyond public accommodation issues and cover employment as well. He even demanded that unions as well as employers be subject to the antidiscrimination requirements of the law.

At the same time, however, Meany insisted that seniority, precious to labor, be protected. The law provided that if a union had negotiated a "bona fide" seniority system—one without discriminatory intent—it would not, even if it happened to adversely affect protected classes, violate the act.

Implementation of the act soon led to friction within the coalition. To the extent that affirmative action succeeded in gaining new positions for minorities and women, white males who perceived themselves to be disadvantaged complained of reverse discrimination. Collective bargaining agreements with departmental seniority systems often locked black and female workers into low-level jobs and effectively barred them from advancement. These problems occurred in an essentially expanding economy, and progressive labor leaders worked with sensitive civil rights advocates to resolve them through such creative mechanisms as the steel industry consent agreement, which integrated departmental seniority lists into plant-wide systems. The harder problems of the post–Vietnam recession—when seniority-based layoffs would place unbearable strains on the coalition—were yet to come.

Concern for protection of affirmative action's limited gains and for preservation of the coalition led many, including the husband and wife employment law team of attorneys Alfred and Ruth Blumrosen, to search for solutions to this new American dilemma. As early as 1968 Alfred Blumrosen had lamented: "It is a tragic irony that the success of the labor–liberal–civil rights coalition in passing the Civil Rights Act of 1964 laid a foundation for its own destruction. But there is no doubt that the problems emerging under the statute including the 'seniority problem' have contributed importantly to the breakdown of this coalition. It has achieved no significant legislation since. . . . Yet the liberal-labor coalition has been an important instrument in the adoption of most of the program of modern social legislation under which we live."

Ruth Blumrosen had spent the year following her graduation from college working in the Michigan UI system and, from that experience, was well aware of the fact that the system was structured, in her words, to "pay off layoff." She was persuaded that STC, apart from the protection it could provide for affirmative action, was fundamentally more consistent with the goals of the employment security system than was regular UI since it encouraged employment rather than unemployment.

By 1975 there was a little-noted flurry of activity as several organizations and individuals, including many contributors to this volume, simultaneously began (in many cases independently of each other) to examine seriously the merits of STC. Martin Nemirow circulated within the U.S. Department of Labor what may have been the first comprehensive American analysis of the program.

In March 1975, the National Urban Coalition discussed the STC option in a seminar on "Affirmative Action in a Recessionary Period." Participants included representatives from industry, labor, universities, and foundations, as well as from the major civil rights groups. The first governmental action stressing work sharing, though without any UI subsidy, was the guidelines issued by the New York City Commission on Human Relations chaired by Eleanor Holmes Norton. These guidelines required employers to consider work sharing as an alternative to layoffs that have a disparate impact on protected classes.

In June 1975, New York became the first state to consider an STC bill. However, the measure introduced in the New York State Legislature languished in committee and died an untimely death.

The States Try STC. Ironically, when enactment of the California law in 1978 finally breathed life into STC, civil rights groups and civil rights concerns—despite their prominence in originally moving the issue onto the nation's political agenda—played but a minor role. They were involved to an even lesser extent in the subsequent passage of the Arizona law.

In California the passage in a statewide referendum of Proposition 13 was mistakenly expected to necessitate a massive layoff of public employees. The atmosphere of crisis which the proposition created prompted the hasty adoption of STC as a tool for ameliorating the impact of these layoffs.

The California law was introduced, passed, and signed within two weeks; obviously that could not have happened without considerable prior research. Bill Greene, the California State Senator who introduced the STC legislation, had long been persuaded that STC merited a trial since it encouraged employment rather than unemployment, and he assigned Robert Rosenberg, a staff researcher, to conduct continuing research on the job preservation possibilities of STC. Greene quite literally took advantage of the Proposition 13 crisis. Following the victory, Rosenberg commented, tongue-in-cheek, that it confirmed Rosenberg's Law: "Major social policy changes normally occur in the presence of three conditions: the right idea, the right time, and the wrong reason."

In Arizona it was a major corporation, Motorola, that was most responsible for causing STC to be considered and adopted. Motorola had decided that STC made good business sense because it would permit the preservation of a skilled work force. The Arizona Department of Economic Security, originally skeptical, and the legislature, in the process of responding to Motorola's direct request for a change in the law, subsequently discovered other virtues in the legislation.

Oregon passed its STC law in the throes of a severe recession to cope with one of the highest unemployment rates in the country. After nine months of operation, a survey conducted among both employers and employees found that both groups expressed an extremely high level of satisfaction.

In Canada, work sharing was also introduced in the midst of the recession of the early 1980s. The minimal costs of participating and widespread acceptance by the media and the public encouraged utilization of the program by a variety of businesses.

Federal STC Legislation. In 1982, a coalition of all sectors—women's organizations, labor unions, religious representatives, and civil rights groups—traveled to Washington to lobby and testify for Rep. Patricia Schroeder's federal STC bill. A major recession and an unsympathetic administration, together with a formula that all could support—work sharing *with* unemployment compensation—made it possible for STC legislation to pass in both houses of Congress. Attached to an important omnibus tax bill, there was little question that it would be signed into law by the President—an event that took place on September 3, 1982.

While the success of the program in California, Arizona, and Oregon would seem to assure additional experimentation, the passage of federal legislation by no means ends the debate or guarantees adoption of STC by all states. The federal law (see appendix) is limited in purpose. It merely requires the Department of Labor to "develop model legislative language" and "provide technical assistance to states to assist in developing, enacting, and implementing . . . short-time compensation programs." Since the states are "encouraged to experiment," they are not required but only "encouraged to consider" the following provisions:

- The workweek must be reduced at least 10%.
- STC benefits should be a pro-rata share of regular UI benefits.
- Availability for work and job search tests should not be required.
- The total reduction in hours under STC should be no greater than layoff hours would have been.
- The employer must continue to provide health and retirement benefits as though the workweek had not been reduced.
- Where there is a union, it must consent to the STC plan.
- During the preceding four months the work force must not have been reduced more than 10 percent by layoffs.

This last provision, included at the insistence of the AFL-CIO, is intended to prevent an employer from frustrating one of the major purposes of the law—to encourage continuing employment for the recently hired, particularly women, minorities, and youth.

The federal legislation is sunset legislation and will expire in three years from its adoption in September 1982. By October 1985 the Department of Labor is required to make an evaluation containing recommendations, including specific recommendations for changes in the statistical practices of the department. This is necessary because many federal programs grant benefits to states depending on levels of unemployment, and STC may mask real unemployment by substituting workweek reductions for layoffs. The law mandates that these factors be evaluated:

1. Impact upon trust funds
2. Preservation of jobs, particularly for the newly hired, women, and minorities
3. Extent of layoffs subsequent to STC
4. Effects of varying methods of administration
5. Effect on tax rates

6. Effect on retirement and health benefits
7. Cost-benefit analysis for employees, employers, and communities
8. Cost of administration
9. Other factors

While this volume does not purport to substitute for the Department of Labor evaluation, the following chapters will consider, and hopefully shed some preliminary light on, many of these considerations.

NOTES

1. Work in America Institute, *New Work Schedules for a Changing Society*, a policy study directed by Jerome M. Rosow and Robert Zager (Scarsdale, N.Y.: Work in America Institute, 1981), pp. 90-91.

2. Ibid., pp. 88-105.

3. Ibid., pp. 88-89.

4. Ibid., pp. 89-90.

5. Comment by Sol Hoffman, vice-president, International Ladies' Garment Workers' Union.

6. Ellen Paris, "It Works!" *Forbes Magazine*, 14 March 1983, p. 39.

7. a. Work in America Institute, *New Work Schedules for a Changing Society*. b. Stanley D. Nollen, *New Work Schedules in Practice*. Van Nostrand Reinhold/Work in America Institute Series (New York: Van Nostrand Reinhold, 1982). c. Fred Best, *Worksharing: Issues, Policy Options and Prospects* (Kalamazoo, Mich.: The W. E. Upjohn Institute for Employment Research, 1981). d. Maureen E. McCarthy and Gail S. Rosenberg, with Gary Lefkowitz, *Work Sharing: Case Studies* (Kalamazoo, Mich.: The W. E. Upjohn Institute for Employment Research, 1981).

8. Juan M. Mesa, *Short-Time Working as an Alternative to Layoffs: The Case of Canada and California* (Geneva, Switzerland: International Labour Office, 1982).

2.
Short-Time Compensation:
The Implications for Management

Ramelle MaCoy
Martin J. Morand

Employers cannot choose to grant or withhold from their employees the benefits of protective labor legislation. The minimum wage, premium pay after 40 hours, workers' compensation, occupational safety and health standards, the right to collective bargaining, and Social Security are all established legal rights. Not only can they not be withheld by an employer, they cannot even be bargained away by a union or voluntarily surrendered by an individual worker.

The fact that this is not true of short-time compensation focuses attention on the centrality of management attitudes toward this alternative form of unemployment benefit. The decision to adopt STC is primarily a management decision. STC is not so much protective labor legislation in the classic sense as it is labor-market legislation akin to the Comprehensive Employment and Training Act (CETA) and other laws that offer training subsidies to private employers.

While individual workers and groups of workers are directly affected, nowhere in existing or proposed U.S. legislation is the opportunity to collect this form of partial unemployment benefit conceived of as a right. All of the five state laws already adopted and the bills being considered in other states require that an employer request and receive approval of an STC plan before any worker can become eligible to receive STC benefits. Even in Germany, where a works council may initiate a request to the unemployment insurance system for approval of an STC plan if the employer fails to act, STC in practice is implemented almost exclusively through employer initiative.

This distinction between STC and other labor legislation explains the important roles that have been played by individual businesses such as

Motorola and organizations such as the Committee for Economic Development (CED) in providing political support for the introduction and enactment of STC legislation. Such advocacy is in sharp contrast to the near unanimity of opposition in the business community to unemployment insurance (UI) at its inception and to subsequent increases in benefit levels.

Employers and the Free Market Economy

Employer organizations have historically opposed unemployment compensation as well as other forms of labor legislation that reduce the pressures upon workers to accept any job at any price. Because that ideological posture shapes the political attitude of employers as a class, as opposed to the more pragmatic reaction of some individual employers, this historic opposition to UI merits a brief review.

Opposition to UI Legislation. Employer opposition to the original UI unemployment insurance legislation was summed up succinctly and candidly by Glenn A. Powers of Industrial Relations Counselors in the early 1930s:[1] "Unemployment insurance has acquired [a bad reputation] in America, due to reported European experience. . . . Most of us are dead set against any legislative approach to these problems." H. W. Story, vice-president of Allis-Chalmers Manufacturing Company, adds: "HR 4142 provides that the unemployed shall receive benefits 'as a matter of right.' . . . As soon as society admits this 'right,' how can it logically refuse to grant benefits to all of the unemployed for the duration of unemployment equal to a living wage?"[2]

More fundamental, perhaps, though less frequently voiced, is a management perception that at least some unemployment is a necessary, even salutary, evil. "The stick—the fear of unemployment—is probably the most effective incentive we know for reducing labor turnover or absenteeism," asserts Hilde Behrend.[3]

In the mid-1930s an even more basic value in unemployment was noted: "The strongest human instinct—self-preservation—operates through fear of economic insecurity to produce the incentive for individual productive effort: in other words the incentive to work. The economic security afforded by such evolution (of Unemployment Compensation) would eliminate the fear of economic insecurity and hence the incentive to work. Thus an apathy toward work . . . over a period of time would cause the degeneracy of the masses—the back-bone of this country."[4]

A close look at the contemporary legislative record on STC seems to indicate that such business opposition as there has been has been founded not on the merits of the actual legislation but rather on continuing fundamental misgivings about the nature of unemployment insurance *per se*. It is social insurance of any kind that is suspect, and STC suffers, in the eyes of some conservative business and political spokespersons, from guilt by association.

One California Chamber of Commerce official who had expressed personal interest in work sharing declined an invitation to address the First National Conference on Work Sharing Unemployment Insurance with the explanation that his participation would be taken by some of his constituents as evidence that he was soft on the *principle* of opposition to paying people for not working—even as an endorsement of government interference in the economy and private enterprise!

Certainly it is not the idea of work sharing as an alternative to layoffs that is foreign to management. Throughout our industrial history, particularly prior to the New Deal initiation of UI, the hours of labor moved up and down in concert with the business cycle. For some individual entrepreneurs this may have been primarily a matter of narrow economic self-interest, a way of hoarding a trained labor force. But it was also part of a more profound interest in reducing turnover—not only to limit its costs but to increase worker loyalty and work-force stability. Spreading the work through temporary reductions in hours was akin to "pensions, stock options, insurance, home financing and other benefits . . . awarded for long service with the company."[5] Labor mobility was a hindrance to the expanding capitalist system, and work sharing was a strategy instituted to overcome crippling turnover rates. In 1912–13, for example, Ford Motor Company experienced a turnover rate of 415 percent in contrast to a 1969 rate of 25 percent.[6]

Even seniority, so often taken to be synonymous with unionism, was a part of the package of incentives and benefits proffered by a host of burgeoning businesses. The fact that company seniority was sometimes awarded selectively, with supervisory judgments as to loyalty and efficiency frequently carrying greater weight than length of service, does not deny the concept but rather is the exception that proves that a rule existed.

Work Sharing in the 1930s. The corporate community not only practiced work sharing (or "work spreading" as it was commonly called in the early 1930s) out of individual self-interest but also in response to patriotic and humanitarian admonishments from both the government and its own business associations. Keeping more workers than were actually needed on the payroll created some small administrative burden and expense, but in that prefringe benefit era when labor was paid only for hours worked or pieces produced, this was a small price compared to the potential costs of political and social unrest. The realization that such unrest might lead to pressure for welfare legislation that would be reflected in higher taxes was a major incentive for avoiding layoff.

Toward this end the Hoover administration created the President's Organization on Unemployment Relief, which occupied itself almost exclusively with encouraging (through a process later to become known as jawboning) and applauding business efforts through work spreading to relieve the pressure on government to "do something."

Success was evidenced in the results of a survey of some 25,000 companies with 1929 capitalizations of $100,000 or more. The President's Organization, summarizing survey data, reported:

> The results of this investigation show how industry and business in their respective branches have spread employment, and indicate where the possibilities for further spreading are most promising. . . . Reporting companies, representing all sizes and practically every type of industry and business, in 1929 employed 3,475,870 persons at a weekly payroll of $104,461,727. During the payroll period ending nearest March 15, 1932, these companies employed 2,547,901 persons at a weekly payroll amounting to $60,626,129. This represented a decrease of 26.7 percent in employment and of 42 percent in payroll.
>
> On March 15, 1932, of those employed, 1,428,116 (or 56.1%) were on part-time. These part-time workers were employed, on the average, 58.7 percent of full time. . . .
>
> "Reduced days per week" was the method most commonly used for spreading or increasing employment. . . . An analysis . . . shows a wide variation in the extent to which work has been spread.[7]

One of the companies that responded to the survey remarked:

> Owing to the unusual situation that has confronted us the past two or three years, we have found it necessary to reduce the number of hours some of our departments are operating to a point where the earnings of employees in departments so affected are hardly sufficient to enable them to meet living expenses. We have even gone so far as to delay putting into operation equipment that would reduce our costs very materially and that at the same time would throw some of our employees out of work. At the present time we are limiting all of our employees, with exception of those on the salaried payroll, to 30 hours per week, and it has been unnecessary for us to hire extra employees even after making this maximum 30-hour weekly schedule effective.[8]

The President's Organization report continued:

> The methods of spreading or increasing employment used [were] reduced days per week; reduced hours per day; shorter shifts in continuous operations; alternating shifts or individuals; [and] rotation of days off. . . .
>
> This survey shows large proportions of our industrial establishments utilize the spreading of work for the maintenance of employment. . . . Such in-

dustries could be approached and encouraged to increase employment where possible, but perhaps the most effective means of adding to present employment should be for each community to make a check of the possibilities with its own industries and businesses.[9]

This historic business approval and practice of work sharing has not been translated into automatic support for contemporary efforts to adopt STC legislation for several readily apparent reasons. In the first place, the fact that STC provides a monetary benefit for the hours reduced—pay for time not worked—casts a different ideological light on the issue. More important, certainly in economic terms, is the expense of maintaining today's high-cost fringe-benefit packages for a larger than necessary number of workers.

Most important, perhaps, is the fact that during the intervening fifty years the adoption of unemployment insurance has rendered layoffs less traumatic and more acceptable than they were in the 1930s when a layoff almost certainly consigned a worker to destitution. The vast majority of personnel managers in America today have no memory of pre-UI days and no driving compulsion to search for any alternative to layoff. It has only been in the high unemployment era of the early 1980s that layoff has been widely feared as anything worse than a temporary interruption in work cushioned by unemployment insurance. It is these fears and the perception that high unemployment rates have become a "normal" part of our economic life that have fueled widespread renewed interest in work sharing.

Although almost certain to be dismissed by an individual employer, a consideration that should be weighed by the employer community is the social costs of layoff. As they evaluate the appropriate place of STC in our socioeconomic scheme, employer groups should attempt an assessment of the long-term and diffuse, but inevitable, costs to business of the crime, delinquency and family breakdown, emotional instability and educational failure, and alcoholism and drug abuse that are directly linked to unemployment. Alcoholism and drug abuse, in particular, are likely to be reflected in an increased incidence of illness that, in turn, will be directly reflected in the cost of some employer's experience-rated health insurance (the conditions are likely to remain untreated and their long-term health consequences unnoticed during the period of unemployment). Employer groups that are fond of pointing out that there is no such thing as a free lunch are right. The cost of unemployment must be paid, frequently by business.

While the employer community tends to resist both legislated and negotiated reductions in the workweek, the enlightened employer, well aware that the organization seldom gets 40 hours of productive labor in the standard 40-hour workweek, will be keenly interested in determining the optimally efficient workweek. While it is generally known that a reduction in hours tends to increase production per hour, what is the point of diminishing returns at a given time in our technological evolution? Almost no evidence is available on this subject. Certainly the long historic trend of workweek reductions as science and industry have progressed has not been sustained in the post-World War II period, possibly because the Fair Labor Standards Act, collective bargaining, and UI have combined to make the 40-hour week seem an absolute norm. Experience with workweek reductions under STC would very rapidly provide a mass of empirical data from which the effects of varying reductions in hours could be determined in different industries, jobs, and circumstances.

UI as a "Fact of Life." When UI first became a fact of life to be dealt with rather than an abstraction to be debated, several new management reactions were manifested. In the legislative arena these took the form of successful pressure for experience rating; at the workplace there was occasionally an effort to schedule work-time reductions and characterize discharges in such a fashion as to deprive workers of UI in order to protect a firm's experience rating—compensation cost control became a function of the personnel office. On the contrary, some employers, either on their own initiative or in cooperation with the union, arranged work schedules, in effect, to protect and preserve their own work force by assuring maximum eligibility for UI. In a panel discussion on "Handling Layoff Problems" conducted by the American Management Associations[10] the following exchange occurred:

Q. (L. Clayton Hill, professor of industrial relations, University of Michigan) "How popular with employees is the alternative-week plan of sharing the work? I mean particularly [with] those workers who would be entitled to full employment on the basis of length of service?"

A. (Solomon Barkin, research director, Textile Workers' Union of America) "When unemployment insurance entered the picture, there was a very definite trend in the direction of adjusting the workweek of retained employees to the requirements for unemployment insurance qualification. As a result we now have in many of our plants a combination arrangement made to suit local insurance needs."

With UI accepted as a normal business cost, some employers, particularly those in seasonal industries that routinely experience wide swings in their need for labor, found ways to match their own scheduling variations to the requirements of the UI rules. The most widespread use of this technique may have occurred in the garment industry, where workers in the slow season would be given alternating weeks of work and layoff in order to qualify. For such employers, STC would do no more than provide an administrative alternative for accomplishing, perhaps more conveniently, what they are already practicing.

It is precisely such practices on the part of employers who contribute fewer dollars to the UI fund than their employees draw in benefits that have led to surtaxes for negative-balance employers in the STC laws of California, Oregon, Arizona, and Florida. Washington has no such surtax provision because its UI does not call for experience rating. Similar surtax provisions are contained in the pending STC legislation of other states and in the proposed federal model law. Employers who "pay their own way" have long objected to the social insurance aspect of UI's pooled funds, and the STC surtax provisions are the direct result of their complaints.

However, in the absence of a full employment economy, it is questionable whether the elimination of such "subsidies" from the UI fund would effect the savings anticipated by the protesters. The earnings of workers in some of these seasonal industries are frequently so low that their annual incomes would be below poverty levels if not for unemployment benefits. It must, therefore, be considered whether society as a whole (including those protesting employers whose roles are so central to the production of the goods, services, employment, and taxes that sustain the society) would be better served if those seasonal industries were replaced by higher levels of unemployment and the social welfare costs and reduced demand associated therewith.

In any event, it is important that where surtaxes are introduced, they are not so punitive as to prevent utilization of an STC system that society has determined socially and economically useful. Surcharges should be designed to recapture for the fund only the additional costs to the fund caused by the use of STC rather than regular UI.

Employers and Short-Time Compensation

Despite the key and crucial role that management, because of the direction STC legislation is taking, will play in the application of STC if the concept is widely adopted, an even more important management role will be in public policy development. Clearly the leading spokesperson for the positive management policy of encouraging STC is Frank Schiff, vice-president and director of research of the prestigious Committee for Economic Development. On July 24, 1975—long before the recession of the 1980s and double-digit unemployment had ignited widespread interest in STC—Schiff told the Conference on the Economy of the House Democratic Steering and Policy Committee that "problems could often be avoided and full-time unemployment would be reduced if greater incentives for share-the-work arrangements were provided through appropriate changes in the unemployment insurance laws."[11]

Schiff's statement was made with specific reference to "increasing problems for business and unions because of charges of reverse discrimination when layoffs are not based on seniority" and of charges of discrimination by recently hired women and minorities when they are. Subsequent policy statements from the CED, however, have not only looked at the STC question from an affirmative action standpoint, but also in terms of the economy as a whole *and* the costs and benefits to the individual employer. Since it captures so well the essence of the business rationale for utilizing STC, we quote at length from a CED statement of January 1978:

> We believe, however, that there is need for a harder look at an additional approach: namely, more systematic efforts to minimize outright layoffs and forced idleness during recessions.
>
> First, business firms should, wherever feasible, fully explore alternatives to outright layoffs when their sales volume is reduced as a result of recessions. The scope of such alternatives will, of course, vary greatly for different companies and in different situations. Firms should not be asked to retain employees where this runs counter to the companies' longer-term productivity. However, we believe that a careful assessment of longer-term costs and benefits would uncover a larger number of instances in which the companies would benefit from a reduced number of dismissals.
>
> Without a careful cost-benefit analysis, companies may underestimate the extent to which outright layoffs in recessions are now often more costly or represent less of a saving than in the past, especially where increased training costs have substantially added to the capital value of current employees and where the cost of searching for qualified new employees is high. Conflicts between seniority and equal employment rights have also increased the costs of layoffs, resulting in extra legal expenses as well as morale problems. At the same time, the savings from layoff have often been reduced because of increased company contributions to unemployment compensation and supplementary benefit payments.
>
> . . . we recommend active exploration of possible legal and administrative changes to facilitate work sharing as an alternative to cyclical layoffs in cases where such a solution is desired by both management and labor. Work sharing can be important in reducing the uneven burdens that now fall on a limited number of employees with the least seniority during recessions. Furthermore, because work sharing avoids layoffs altogether, it can be particularly useful in preventing conflicts between seniority and equal employment considerations during recessions.
>
> The needed administrative or statutory changes might include allowing payment of unemployment insurance for single days in cases in which work schedules have been cut to four days. . . . It should be emphasized that our recommendations relate to work sharing only as an alternative to cyclical

layoffs; consideration of the issues involved in wider use of work sharing and shorter hours over the longer term is beyond the scope of this statement. Moreover, our comments concern the kind of work sharing that produces four days' work for four days' pay, not four days' work for five days' pay.[12]

The essence of management is decision making, but decisions can be made only when options exist. Whatever its other virtues, STC makes available to management an option previously unavailable: to reduce labor costs either through layoff or work sharing.

American experience to date is too limited and the data thus far accumulated not evaluated thoroughly enough to provide managers any certain guideposts to the best decision. Noah Meltz, Frank Reid, and Gerald Swartz in *Sharing the Work: An Analysis of the Issues in Worksharing and Jobsharing* have made an effort at developing quantitative economic formulae for analyzing the costs and benefits of work sharing, but since both costs and benefits are specific to each firm, these issues will be examined here conceptually rather than quantitatively.[13]

Fringe-Benefit Costs

The single major cost to the employer of choosing STC is the maintenance of fringe benefits, and the most costly fringe to maintain is almost always health insurance. Except in cases where the employer pays for health coverage through a per-hour, per-ton, or per-dollar of earnings contribution (usually to an employer-union welfare fund), health coverage is usually paid on a per capita basis, and maintaining extra workers on the payroll adds a heavy burden at a time when business is bad.

This problem is particularly acute in the United States. American companies with plants in Canada that have participated in that country's STC program find that maintaining health insurance coverage for retained employees is far cheaper, since coverage there need only supplement Canada's national health insurance, which covers employed and unemployed alike. In Germany the unemployment insurance fund subsidizes a portion of the extra fringe costs of carrying more workers on the payroll because of STC.

Other fringes with major cost implications include all forms of paid time off (vacations, holidays, and sick leave) and pensions. A company with a contractual or personnel policy obligation to provide two weeks of paid vacation at the regular hourly rate after one year is probably going to pay two weeks to all those kept on the payroll even though they may have spent several weeks at less than full time. On the other hand, there are contracts and practices under which vacation is paid as a percentage of earnings and in these cases STC would not affect labor costs. The cost implication of sick days and holidays will similarly depend on the specifics of the entitlements, but in most cases extra personnel will mean at least some additional costs.

Pensions are more complex. In cases where employer contributions to a pension fund are based on earnings or hours worked, the use of work sharing should have no impact on costs. However, in the case of defined benefits, where employees are guaranteed a certain level of benefits based on earnings for a specified period or for years of service, work sharing could have the long-term effect of increasing employer costs, but so could temporary layoffs under certain circumstances.

Legislated fringes such as workers' compensation, unemployment insurance, and Social Security—all of which are paid as a percentage of earnings—will, in practice, probably cost approximately the same with STC as with layoff. However, there may be exceptions that will depend on the cap on taxable earnings, the mix in earning levels among those laid off and those retained, and the timing and duration of layoffs. If the earnings of 100 workers, for instance, have all exceeded the taxable base for a given fringe for the year in question, then keeping all of them on the payroll under a work-sharing plan will be no more expensive for that particular fringe than keeping 80 and laying off 20. In some situations, depending on the timing and duration of layoff, the choice to use STC may adversely affect cash flow but not the ultimate annual cost.

The Cost of Labor Turnover

There is little question or disagreement that the maintenance of fringes is the largest single cost of STC to the employer. It is similarly accepted—and largely confirmed by the California data—that the elimination of layoff-related turnover cost is the greatest saving for most employers.[14] Unfortunately, while the calculation of the STC cost impact on fringes may be complex, the calculation of the cost of turnover in most businesses is even less of a business science. While there is no disagreement that the costs of turnover to business and industry are generally staggering, individual businesses and industries have been curiously lethargic in attempting any systematic accumulation of data upon which to base reliable estimates of the actual cost to them of turnover. Indeed, the single greatest contribution that the availability of the STC option may ultimately bring to the business community may be to force managers, in order to make cost-benefit analyses, to more adequately and realistically estimate and weigh the true cost of turnover.

The failure of American management to give sufficient attention to the cost of turnover has been frequently noted: "In general . . . management regards labor turnover as costly and wasteful. Unfortunately, the information usually given to top management about the extent and cost of turnover within the plant is either nil or very meager and, therefore, possibly misleading."[15]

It has been similarly pointed out that ". . . managers frequently receive periodic reports on spoilage or wastage, accident frequencies, even absenteeism—but not labor turnover. This is a curious lack, in view of the fact that the basic causes for changes in the reports relating to spoilage, accidents, and cost of productivity are partly due to the plant's labor turnover."[16]

The accuracy of the cost estimates that are made is questionable because of the lack of consistency, says the same study.[17] It reports (from a survey of corporate controllers on "Estimated Range of Costs to Recruit, Process, Orient and Train to Average Production Level") that sums ranging from $50 to $2,000 are estimated as necessary to fill the simplest clerical job; in clerical jobs requiring higher skills estimates were even wilder, ranging from $250 to $7,000; for journeyman machinists the range ran from $125 to $5,000. From such a data base, management decisions on layoff cannot be made with any degree of intelligence.

Clearly, the choice of STC or layoff will be affected by the value attached to the labor and related costs of layoff-generated turnover. It is equally clear that if management has not been sufficiently motivated in the past to attempt the accurate assessment of the cost of layoff because it was viewed as the "given" or "accepted" means of reducing work time, the availability of the STC option may provide the motivation to gather accurate layoff cost data.

Another potential contribution of STC to individual businesses and economic planning may lie in the cost-benefit studies of this new public policy. Beyond the congressionally mandated evaluation study to be made by the U.S. Department of Labor and similar studies likely to be undertaken by some of the states, academic research is already being initiated. Two of the contributors to this volume are completing doctoral dissertations on the California project.

The very process of assessing turnover costs will be beneficial. Once we move beyond the obvious costs of searching for qualified new employees, such as advertising, aptitude testing, interviewing, psychological testing, reference checking, medical examinations, security and credit investigations, and applicant travel expenses, we turn to the task of identifying on-the-job costs. These include: putting the employee on the job, company badge, safety glasses, general indoctrination and training, formal job training programs, and break-in costs that might include the higher cost of productivity, increased cost of supervision, higher inspection costs, increased maintenance and depreciation costs, and higher accident costs.

One interesting aspect of this listing of productivity costs is what is not included on any of the turnover cost assessment lists we have reviewed. At best, the typical effort to evaluate turnover costs attempts to measure the time (and money) needed to get the new worker up to speed. But, as knowledgeable commentators have pointed out, work is usually performed by groups rather than by individuals, and the layoff, recall, or new hire of a single worker—because of bumping, transfers, or simple additions to the work group—may have the effect of disrupting and causing great damage to entire work teams, formal or informal, within the plant.[18] If we accept the persuasive premise that productivity is the product of teamwork, then the mobility and turnover that characterize the American labor market may explain much of the widely advertised American competitive disadvantage *vis à vis* more stable foreign labor forces.

Another "productivity" factor much discussed but seldom significantly considered in lay-off decisions is morale. Since the relationship between job security and morale and loyalty is detailed in chapter 12, we will not review it here, but it is clearly a cost-benefit consideration that will be weighed by the thoughtful manager.

Administrative Costs of STC

To the extent that STC requires employers to file applications and process forms to gain approval of a work-sharing plan and establish worker eligibility, it may add administrative costs. The way in which the law is administered, however, can materially affect both convenience and cost. One large California employer argued logically and unsuccessfully that the Employment Development Department could take all the information needed to pay each worker the appropriate benefit from a computer printout prepared for other purposes. California officials were not willing to accept the printout, but such a procedure is routinely followed in Germany with mutual convenience and savings to both employers and the employment security offices.

There is probably a natural conflict between the employers' desire for a program that combines ease of administration with maximum flexibility of utilization and the government agencies' desire to enforce a degree of uniformity and keep its administration as efficient as possible. There has been a tendency for state agencies to hew as closely as possible to policies and procedures already in place for regular UI.

Few firms confronted with the necessity of laying off production workers find it feasible to lay off proportionate numbers of clerical, managerial, and supervisory personnel. In a plant of 200 employees, for example, the layoff of 40 production workers will not reduce the need for a foreman or payroll clerk. If the firm opts for STC, however—particularly if the hours reduction should take the form of a four-day week—it would be entirely possible for the firm's supervisory, managerial, and executive employees to share the reduction with the production workers and thus effect additional savings. Such a procedure would also eliminate the bitterness frequently felt by production workers because they are so frequently forced to bear the entire burden of the need for a reduction in labor costs.

Absenteeism, Accidents, and Waste

The relationship between productivity and the length of the workweek is generally accepted to be an inverse one, but the length of the optimally productive workweek for various industries and jobs has not been established. Available data, however, indicate that the most productive workweek is most certainly less than 40 hours, and the preliminary production data from work-sharing situations seems to confirm this. But it may prove difficult to ascertain how productivity increases measured in work-sharing situations should be apportioned between increases attributable to the shorter workweek and increases due to higher worker morale.

Specific aspects of productivity have been positively identified as being more economical and efficient during periods of work sharing. The introduction of work sharing at the McCreary Tire Company (achieved in the absence of STC under regular UI by the use of rotational layoffs) resulted in "increased productivity . . . on a pounds-per-worker-hour basis . . . a drop in the absenteeism rate, a dramatic reduction in the number of injuries, and a significant decline in the percentage of scrap waste generated." The reality of the relationship between morale and productivity was confirmed by the author's findings: "The fact that many of these gains specific to the . . . [work sharing] period tended to continue beyond that period suggests the true significance of the work-sharing experiment: what really happened was not so much a change in hours scheduled as a change in the entire labor-management relationship."[19]

Absenteeism is normally a voluntary decision not to go to work because the worker is ill, tired, emotionally upset, or must attend to personal matters. Some planned absences—for medical examinations, legal consultations, dental appointments, and so on—are eliminated when scheduled time off during the regular workweek is available under work sharing.

STC and the Decision to Lay Off

Since the decision to lay off is often more painful for management to make than a decision to share the work, it has been speculated that STC may facilitate an earlier and thus more economical response to depressed market conditions. While this seems logical, a countervailing consideration has surfaced in one company interview. The company president explained that once an announcement has been made that work will be shared by reducing hours to 32 for a 20-week period, it is difficult for the company to make any additional reductions or layoffs during that period. The announcement is apt to be taken by employees as a commitment made in exchange for their sacrifice in accepting the reduced hours, and any deviation from it may be perceived as dirty pool.

From the employer's viewpoint, one disadvantage of maintaining the work force through utilization of STC is that it may tend to deprive supervisors of the opportunity to weed out undesirable or unproductive workers through selective layoffs. In practice the situation will be different for union and nonunion operations and, depending on company policies and practices, be of greater or lesser importance in a nonunionized company.

Under a union contract with a strongly enforced discharge clause, the only available method for eliminating undesirable workers may be through periodic layoffs of sufficient length to exhaust recall rights. Even during short layoffs, the possibility exists that unsatisfactory workers, who may be presumed to find little satisfaction in their jobs, will find employment elsewhere. The concomitant danger is that the better workers will be more likely to seek and find alternate employment.

Theoretically, the differences in union and nonunion operations may appear slight, since many or most nonunion employers, as a matter of policy and in the interest of morale, follow the identical layoff in inverse order of seniority procedure required by most union contracts. In practice, however, the situation may be radically different, since the nonunion employer is almost totally free, as the union employer is not, to eliminate undesirable workers through selective recall.

It has been mentioned that the practice of work sharing tends to increase or at least maintain productivity during periods of reduced hours because of improved morale, the greater productivity associated with shorter hours, and the elimination of the disruptions and inefficiencies caused by the bumping and reassignments normally necessitated by layoffs. Even more important may be the capability work sharing gives management to respond to even temporary or sporadic increases in demand by quickly scheduling additional hours.

Returning to full production from the layoff mode is quite a different matter and may not be a feasible response to temporary improvements in demand. Some of the workers on layoff will have found other jobs for which replacements will have to be recruited and trained, and the laid-off workers who do return to work will cause the same disruptions occasioned by their departure as workers are shifted and reassigned.

STC: Not for All Employers at All Times

Since clearly the desire to hold a skilled work force in expectation of future needs is the primary advantage of STC for most employers, it is equally clear that layoff is the preferred choice for the employer who expects, because of choice or circumstances, to permanently reduce the work force. An August 10, 1982, internal U.S. Steel Corporation memo to its general superintendents illustrates the point: "As we view the 1983 Business Plan, it projects a significant increase in operations and, therefore, a significant increase in employment levels. . . . We must avoid, whenever possible, recalling laid off employees since by doing so we would re-entitle them to benefits that might otherwise have run out. . . . The judicious use of overtime is preferable."[20]

The U.S. Steel policy calls attention to several other circumstances that might affect a particular firm's decision to adopt or reject STC:

1. *Mobility of capital.* While a service industry or public utility may feel an obligation to preserve employment for workers in the community it serves, a corporation that has an opportunity to increase profits by a transfer of operations to a location with cheaper labor or other business advantages is not likely to be equally concerned about sustaining its employees. While there are differences dependent on the nature of the industry, there are also differences among firms within a given industry. For instance, other steel companies have been concerned to preserve their labor force because of policy decisions to continue production at existing locations.

2. *Employee ownership.* In the few but increasing number of cases where employees themselves are the sole or principal owners of a firm, STC is obviously the option of choice. Most such employee-stockholders invested in the company in search of job security, not dividends, and in such situations it is extraordinarily difficult to lay off a stockholder.

3. *Other forms of employer-employee "partnership."* A company that has stressed, for whatever reasons, that employees are "partners" or members of the company "family" may find layoffs particularly damaging to morale and, therefore, choose STC instead. Similarly, an employer who has solicited or needs employee-union cooperation in lobbying programs

(e.g., import limitations) will want to give extra consideration to employee loyalty and employment stability.

4.*Utility companies.* The public utility is somewhat like the public employer—only more so. As a community-based service operating under a government monopoly, the public utility is subject to rate reviews by political entities and can be severely damaged or mightily assisted by its public image and the morale of its employees.

5. *Affirmative action employers.* Since newly hired employees are generally the first to go in a layoff, firms, public or private, that have made large investments in affirmative action programs will find particular value in work sharing. Work sharing makes it possible to retain recently hired employees without repeating the expensive process of recruitment and training. In general, STC is chosen because an employer really values its employees.

Some Special Considerations for the Public-Sector Manager

While the first American short-time compensation law in California was intended specifically for application to public employers and employees, there has been virtually no usage of STC in the public sector in California, Arizona, or Oregon, despite the fact that there are even more compelling reasons for its use by public than by private employers. And since public employers are exempt from the requirements of the UI system and may merely reimburse the trust funds dollar-for-dollar for benefits paid out, there is no reason that many public employers could not directly implement the practice of subsidized work sharing even in the absence of an STC law.

There are two fundamental reasons that may explain this failure on the part of public employers to utilize STC:

1. The public sector has minimal experience with layoff, and public employers typically have little understanding or appreciation of the true costs of layoff. An interesting but little noted 1978 study commissioned by the State of New York and its Civil Service Employees Association concluded that layoffs would actually cost the state more than they would save during the first year of implementation because of the direct reimbursement of unemployment benefits, a combination of the kind of costs that layoffs generate for any employer, and the special costs that accrue to the public employer in its responsibility to protect the health and safety of its citizens.

2. The public employer's personnel policies are frequently molded by political as well as economic considerations, and any long-range planning that extends beyond the date of the next election is likely to be assigned

a very low priority. Since the principal advantages of STC are long range (particularly the reduction of hiring and training costs), employers in the public sector are willing to balance this year's budget and pass on the problem to future administrations.

When unemployment compensation coverage was first broadened to include public employees, UI premiums were not required of public employers. An alternative was offered in the form of a pay-as-you-go plan, the method of payment chosen by most public agencies. Since these agencies had no experience with layoffs, they did not expect to pay UI premiums. As a result, such employers have not built up any reserve in the fund and must reimburse the fund for benefits as they are paid. This reimbursement cost inevitably triggers additional budgetary impacts and further layoffs.

Nonetheless, there are several considerations that encourage the use of STC in the public sector even more than in the private:

1. Layoffs that may cause poor public relations in the private sector may have even more serious negative political reactions in the public sector from the employer's point of view.
2. Layoffs under civil service procedures or public employee union contracts are often more complicated and costly in their bumping and transfer provisions.
3. The affirmative action obligations of the public employer are generally more stringent and complex than those of the private employer. Layoffs that affect women, minorities, the handicapped, or veterans negatively may lead to costly litigation as well as political repercussions.
4. Layoffs lead to the immediate loss of tax revenues.
5. While all employers ultimately share in the societal costs of layoffs, the cost to the public employer—as the provider of welfare and social services—is immediate and direct.

Thus far, STC has avoided the acquisition of any political label. It has been sponsored and supported by both Democrats and Republicans, conservatives and liberals. Whether it is more accurately defined as a worker benefit or a management tool is unclear and perhaps irrelevant. It may ultimately turn out to have been a first step in the evolution of a rational employment policy in a period of reduced employment opportunities.

Work Schedules with STC

The manner in which a reduction in hours will be achieved under STC will vary with the nature and requirements of the employer's business.

The 24-hour operation will require one form of scheduling adjustment, the service enterprise that must be open to the public five and sometimes seven days a week another, and the manufacturing operation yet another.

Many manufacturers will find it most economical to simply reduce the workweek by one or two days and thus effect additional savings in heat, electricity, and other overhead costs. Such an arrangement, with its three- or four-day weekend, is also likely to be particularly appealing to employees.

One California firm alternated a regular five-day week with a week that consisted of three days of work and two days of STC benefits, thus doubling the period during which it could maintain a reduced production schedule with STC—something that would not have been possible had it simply gone to a four-day workweek. The plan had the additional advantage of giving all employees a four-day weekend every other week.

It is important to emphasize that STC does not compel any one reduction schedule for all of a firm's employees. It is possible that there could be one group of employees scheduled for a four-day, 32-hour workweek while another work unit (any group of employees designated by the employer to participate in a particular worktime reduction schedule) could be assigned to a three-day, 24-hour (or three-day, 30-hour) schedule. Still other employees might be continued on a regular 40-hour schedule in response to business necessities or employee preferences. With proper planning and employee consultation, both flexitime and voluntarism may be combined with STC to the mutual advantage of employer and employees. Even though none of the five existing American STC laws require the approval of the work force where it is not unionized (as do the German and Canadian programs), the employer will probably find that prior consultation will enhance the morale and productivity benefits that prompted the choice of STC in the first place.

Before employers in all states except those with STC legislation confront, or can confront, the option of choosing between layoff and STC, they must decide to accept or decline the opportunity to play a role in the legislative process that is considering the adoption of STC programs in those states. Even in states where no actual legislation has been introduced, an employer could take the initiative in urging the introduction of STC legislation, as Motorola did in Arizona.

There are obviously sound business reasons for some employers to favor adoption of STC. What may not be obvious is that there are also legitimate business reasons for them to be active in the legislative process. Through such involvement they have the opportunity to shape the ultimate laws for ease of administration and to meet, as nearly as possible, their own business needs.

Examples of the ways in which legislative language and administrative procedures may affect the practical operation of the program are available from the California, Oregon, and Arizona experiences. In California and Oregon, "moonlighting" earnings by STC claimants are offset against STC benefits. It seemed a logical decision in those states to treat moonlighting exactly the same for STC benefits as it is treated for regular UI benefits. In Arizona, however, it was decided—with equal and perhaps more practical logic—that moonlighting earnings were irrelevant to a program in which the standard test of availability for other employment was deliberately rejected.

The importance of the issue to the employer lies in the fact that no firm is likely to want to impose STC over the strenuous objections of its workers. To the extent that many workers regularly earn moonlighting income by repairing cars or TV sets, pumping gas, or whatever, they are certain to be bitter if fellow workers receive STC benefits from which they are barred because of such income.

A similar problem, ultimately solved through legislative action, existed in California because the law provided for benefits only in weeks in which *work* time was reduced. Paid holidays were defined as nonwork time, and the income received from the union by union stewards for released time spent handling grievances was defined as moonlighting income.

A problem arose when a union steward was denied STC benefits during a week in which paid holidays, scheduled STC days off, and time spent handling grievances combined in such a manner that he was deemed to have performed no work. At the same time, his income from the holiday pay and released union time was too great to permit eligibility for regular UI.

In this and other instances, problems were created because STC was treated too much like regular UI, and the distinct and different purposes of STC were ignored. Perhaps more than any other group, employers can be knowledgeable and effective in insisting that STC laws and administrative rules and procedures be relevant to the purposes of the program.

It has been asserted frequently that a management choice of STC rather than layoff might be made in part because it means a reduction in hours for all employees, including the more senior and presumably higher-paid workers, rather than the layoff of only the junior and presumably lower-paid workers.

While some firms may find these assumptions confirmed in their workplace, there is no evidence that a significant wage differential *based on length of service* is to be found within job categories in American industry. It is almost certainly not the case for a large number of jobs where an incentive pay system is in effect. Layoffs are often made by job category or department rather than by plant-wide seniority.

The assumption that longer service means higher earnings is so widely held in academic circles[21] that the cost-benefit analysis in the California study[22] built this assumption into the model and, therefore, inevitably concluded that the weekly benefit rate under STC could be slightly higher ($1.02 per week) than under regular UI.

We have not been able to find any empirical research that validates this theory or the claim that flows from it—that STC will be more costly to the UI fund. Our own limited research and extensive experience with the American system of labor relations also leads us to question it. Even though it is true that pay for the same job varies significantly between entry-level and senior personnel in government and academic employment (the experience upon which the theory rests), it is not true of American workplaces, particularly in the range between the minimum wage and something under $10 per hour (at which point maximum UI rates are received anyway). In contrast with Japan, China, and perhaps some other economies, the norm in America is payment for the job description or job performance rather than for such individual characteristics as age or employment tenure. To the extent that American private enterprise rewards and encourages continuity of employment, it usually does so in the form of vacations, pensions, and other fringe emoluments. Evidence confirming this experience was obtained in two forms in the course of a study of STC in California.[23]

Within the same plant, a large manufacturer had two bargaining units covered by contracts with the International Association of Machinists (IAM) and the International Brotherhood of Electrical Workers (IBEW) who adopted STC. The personnel director had the payroll department do wage cost simulations for both layoff and STC. Average wage rates were computed for both before and after layoff, and the differences were 5 percent in the production (IAM) unit and 1½ percent in the craft (IBEW) unit. But because all the workers in both cases earned above the level necessary to yield the maximum UI benefit, the impact on the fund would be nil. In addition, contracts from the unions participating in the California study were analyzed in terms of wage schedules in relation to the layoff system. A strict layoff by plantwide seniority system was in effect in only one case, and in that case wage rates throughout the plant varied by only slightly over 5 percent.[24]

In all other contracts, layoff was by occupational category or by department. In most job classifications there was a hiring/probation rate and a job rate attained usually within several months of hire. The departmental seniority procedures did permit the retention of more senior workers within the department but only if they bumped down to lower-rated jobs and were paid at the lower rate.

While an occasional employer may find wage differentials to be an important consideration in deciding to opt for STC rather than layoff, we doubt that such will often be the case. Even if the assumptions of the wage differential between junior and senior workers ultimately prove correct and the $1.02 extra benefit cost found in the California study is substantiated, an extra 20 cents per day in costs for STC claimants seems far too trifling to influence consideration of a program with such potential benefits to both employers and employees.

NOTES

1. Glenn A. Powers, *Methods of Minimizing the Effect of Business Depression on the Working Forces*. Personnel Series (New York: American Management Associations, 1931).

2. H. W. Story, "The Wisconsin Unemployment Reserves and Compensation System," in *Proposals and Possibilities of Federal Legislation for Unemployment Compensation*, ed. Bryce Stewart, Personnel Series (New York: American Management Associations, 1935), p. 13.

3. Hilde Behrend, "Absence and Turnover in a Changing Economic Climate," *Human Factor* 27 (2 April 1953): 69-79.

4. H. W. Story, "The Wisconsin Unemployment Reserves and Compensation System," p. 14.

5. David Montgomery and Ronald Schatz, "Facing Layoffs," in *Workers' Control in America*, ed. David Montgomery (Cambridge, England: Cambridge University Press, 1979), p. 140.

6. Ibid., p. 140.

7. William J. Barrett, "Extent and Methods of Spreading Work," *Monthly Labor Review* 35 (September 1932): 489 ff.

8. Ibid., p. 490.

9. Ibid., p. 492.

10. American Management Associations, "Handling Layoff Problems," in *The Practical Meaning of Management Statesmanship*, Personnel Series (New York: American Management Associations, 1949), p. 28.

11. Testimony by Frank Schiff before the Conference on the Economy, House Democratic Steering and Policy Committee. Unpublished transcript.

12. Committee for Economic Development, *Jobs for the Hard-to-Employ: New Directions for Public-Private Partnership* (Washington, D.C.: Committee for Economic Development, 1978), p. 77.

13. Noah Meltz, Frank Reid, and Gerald Swartz, *Sharing the Work: An Analysis of the Issues in Worksharing and Jobsharing* (Toronto, Ont.: University of Toronto Press, 1981).

14. State of California Health and Welfare Agency, *California Shared Work Unemployment Insurance Evaluation* (Sacramento, Calif.: State of California Health and Welfare Agency, 1982).

15. Frederic Gaudet, *Labor Turnover: Calculation and Cost* (New York: American Management Associations, 1960).

16. Ibid., p. 11.

17. Ibid., p. 60.

18. Lester Thurow and Steven Sheffrin, "Estimating the Costs and Benefits of On-the-Job Training," *Economie Appliquee* 30 (1977):507-519.

19. Linda Roberts, "Work Sharing: A Rotational Furlough Plan Saves Jobs at McCreary Tire," *World of Work Report*, February 1982, p. 9.

20. Pittsburgh *Business Times*, 5 September 1982.

21. See, for example, Fred Best, *Work Sharing: Issues, Policy Options, and Prospects* (Kalamazoo, Mich.: The W. E. Upjohn Institute for Employment Research, 1981), and chapters in this volume by Martin Nemirow and John Lammers and Timothy Lockwood.

22. State of California Health and Welfare Agency, *California Shared Work Unemployment Insurance Evaluation*.

23. Martin J. Morand and Donald S. McPherson, "Union Leader Responses to California's Work Sharing Unemployment Insurance Program," paper presented at the First National Conference on Work Sharing Unemployment Insurance, San Francisco, 15 May 1981.

24. Ibid.

3.
A Labor Viewpoint

Ramelle MaCoy
Martin J. Morand

The attitude of workers and their unions is critical to the implementation of short-time compensation (STC). Under the terms of the five state laws thus far adopted as well as the language of the federal legislation urging other states to adopt STC, unions hold absolute veto power over implementation of an STC program in any firm operating under union contract.

The passage of state legislation amending the unemployment insurance (UI) laws—laws that workers and unions usually perceive as *their* most vital benefit from state government—is a highly sensitive, even emotional, issue for labor. The union view of work sharing with a partial unemployment compensation subsidy is therefore critical to any consideration of the future of STC. These views have been and are likely to continue to be an uncertain mixture of opposition and support, suspicion and enthusiasm.

The opposition is deeply rooted in basic experiences and philosophies involving union commitments to full employment as the indispensable foundation for all of labor's goals and to the principle of seniority as the major weapon with which to defend workers against arbitrary or discriminatory workplace decisions.

To the direct question "What does the American labor movement think of work sharing?" the answer would have to be: "Not much." To labor, sharing the work sounds supiciously like sharing the poverty—a proposition in conflict with labor's traditional goal of sharing the wealth.

Work sharing—"workspreading" in the jargon of the time—was used during the Hoover Depression period, unionists recall, to force workers to shoulder the entire burden of an economic condition not of their making. Thus the 1981 endorsement of STC by the AFL-CIO's Executive Council[1] did not signal a retreat from union demands for full employment but rather labor's continuing support of another principle of American trade unionists that was enunciated most memorably by Samuel Gompers: "As long as we have one person seeking work who cannot find it, the hours of work are too long."

The endorsement of STC by state and national AFL-CIO leaders does not represent a retreat from labor's attachment to the principle of seniority as a basic protection. To most modern-day union members and leaders, seniority rights, won for the most part shortly after the Wagner Act legitimized collective bargaining, are sacrosanct.

Increasing labor support for STC—recently in evidence on national, state, and local levels—is an acknowledgment that full employment is simply not an immediately attainable goal. Indeed, the projections by both the Bureau of Labor Statistics and nongovernmental economists are for continuing high levels of unemployment even after the recession of the early 1980s (by whatever definition) ends. As the AFL-CIO statement on short-time compensation put it, "The AFL-CIO has always been in favor of a full employment economy and programs designed to achieve full employment. Unfortunately, many workers face the threat and reality of layoffs. In some circumstances, the existence of short-time compensation may be an alternative that cushions the impact of unemployment."[2]

The American labor movement is very American, very pragmatic. It will come as no surprise to those familiar with the variety and flexibility of American unions as they actually operate that individual unions bargain over wages, hours, and working conditions in different ways and with remarkably variegated results. This is no less true of the approach of unions to STC.

For example, one of the purposes of seniority-based layoffs is to protect the *right* of part of the work force to collect unemployment benefits. Particularly distasteful to unionists—and all workers, for that matter— is the circumstance where work schedules are deliberately manipulated so as to prevent any worker from being eligible to draw even partial unemployment benefits. Senior workers who would normally not be laid off have their hours cut, but none of the workers, senior or junior, receive any unemployment benefits. In this circumstance, workers feel that they are all being cheated out of a hard-won legislated fringe benefit paid for out of their wage package. They see unemployment compensation as a right intended to help workers pay their bills during recessions and thus sustain community purchasing power.

Unions and the Principle of Seniority

Because of the importance that American unionists attach to the principle of seniority—indeed, to some it is almost synonymous with unionism—

it is easy to forget that the practice of dealing with a reduced demand for labor in a given workplace by laying off workers in inverse order of seniority is a phenomenon of relatively recent vintage. The more common practice during most of our industrial history has been to share available work through a reduction in the normal hours of work. Given union controls that prevent arbitrary or discriminatory employment decisions, such as closed shops, hiring halls, and standardized wage rates, sharing was more common than seniority in early union contracts and practices. In 1955, at the time of the passage of the Wagner Act, few collective bargaining agreements outside the railroad industry contained the seniority clauses so familiar in contemporary agreements. Closed shops protected union members and activists; hiring halls gave unions power over allocation of work opportunities (first man on the bench—first referred); and uniform schedules removed an economic incentive for displacement of senior workers.[3]

Indeed the practice of sharing was so common and accepted that, as recently as 1938, a UAW leader in Michigan, Emil Mazey, had to patiently explain the advantage of layoffs over work sharing to UAW members. Mazey used an example to make his point: "In a plant working 24 hours per week with an average wage of one dollar per hour, if we worked one shift, each worker would receive $24 per week. If this work was equally divided by working two shifts, the average income would be $12."[4]

Mazey pointed out that if half the workers continued to work 24 hours per week and earn $24, the other half could be laid off and obtain WPA jobs at $15 per week—giving the entire group average weekly earnings of $19.50 instead of $12 per week. A few months later, when Michigan made its first unemployment compensation payment, Mazey would have substituted UI benefits for WPA jobs.

The use of seniority to avoid social conflict is at least as old as the right of the first-born male to inherit the land in virtually all agrarian societies. The absence of such an accepted procedure would have given rise to multiple dangers, including fratricidal rivalries and the ultimate division of land holdings into parcels of uneconomic size.

In the workplaces of industrialized America the emergence of strict and enforceable seniority rules was due primarily to the almost universal yearning to curb arbitrary and capricious decisions by foremen. The history of the great union organizing drives of the 1930s and 1940s is replete with stories of favoritism by foremen. In order to secure choice work assignments and relative immunity from discharge and layoff, workers curried the favor of their foremen with Christmas and birthday gifts. Workers washed the foreman's car, mowed his lawn, painted his house, and in some instances, sent their wives to perform domestic chores in his household. More intimate favors from women workers are said to have been commonplace.

With layoffs based on "merit"—and with the foreman the sole arbiter of merit—it was not unknown for a planned layoff of 10 percent to create such a flurry of productivity as workers scurried to demonstrate their merit that 20 percent could be laid off instead.

Second in importance perhaps only to the promise of increased wages and benefits, the desire for "fairness" in the workplace was a principal issue on which the great organizing campaigns of the 1930s and 1940s were built. Seniority was the union antidote for arbitrary and capricious personnel decisions, and it was, and is, an antidote with the great virtue of being accepted as "fair" by an overwhelming majority of workers. Seniority is totally objective (in sharp contrast to other criteria for preferment), and even junior workers disadvantaged by seniority are likely to accept it as fair, with the expectation that they will also be its beneficiaries eventually.

Since seniority is so universally accepted, it tends to be the union leader's response to every question requiring the preferment of one member or group of members over another. The typical union contract may contain scores of references to seniority and unions. Although it is seldom completely successful, the contract strives to have seniority exclusively govern order of promotion, shift preference, job bidding, vacation scheduling, and order of layoff. Small wonder that seniority has come to be synonymous with unionism.

Since unionists are likely to see work sharing initially as a threat to seniority, first reactions have tended to be negative. Considerable work sharing in the form of reduced workweeks has regularly been going on, however, even in the absence of STC, and union leaders have naturally taken the position that people on reduced hours should receive a partial UI benefit. The result has sometimes been that union leaders have taken the seemingly ambiguous position of opposing work sharing but favoring STC when there is work sharing.

By taking much of the sting out of work sharing for senior workers, STC, may, in the long run, make it possible and feasible to "share layoffs" as well as share overtime.

The potential advantages of such sharing for unions and union leaders are great. A union's strength is in the numbers of its members, and work sharing maintains union membership and union income (except in cases where dues are a percentage of earnings). Work sharing also avoids the considerable expense and inevitable divisiveness of the disputes, grievances, and occasional legal litigation that accompany layoff.

Work sharing also avoids the increasingly troublesome question of superseniority for union stewards and officers. As layoffs have cut deeper during the severe recession of the 1980s, some union members with many years of seniority have resented seeing union officers with far less seniority continue to work. The justification for superseniority, of course, is that union stewards and officers must be on the job in order to process grievances and administer the contract, which would not be possible were they on layoff. During periods of work sharing, stewards and officers share the layoffs but are on hand when needed for contract administration since they work the same hours as everyone else.

Significantly, the one area where seniority is seldom used is that of overtime. The typical union contract is apt to provide that overtime be shared "equally" or "equitably" by all members of a work group, with senior workers seldom gaining more than a place at the head of the overtime line.

Overtime is shared because it is *possible* to share it—unlike promotions or vacation scheduling. Moreover, no one is disadvantaged or "cheated" by such sharing, since even the most senior workers are not considered to have any "right" to overtime.

During the period since the New Deal, the use of work sharing to deal with temporary declines in demand has decreased, but it has sharply increased, through the use of overtime, to meet temporary or seasonal surges in demand. There is scarcely an employer who does not find it more efficient and economical to respond to temporary surges in demand— whether occasioned by a special order, a rush project, a natural disaster, or a holiday rush—by increasing the hours of existing employees rather than hiring additional workers.

In most cases, there is no way that any employer can immediately increase production except by the use of work sharing through overtime. Except in rare circumstances, there is no way that an employer on a given workday or workweek could increase production through the recruitment, hiring, administrative processing, and training of additional workers. Indeed, it is entirely possible that the sudden introduction of new workers into a work group could have the immediate effect of reducing production. But overtime—another name for the sharing of surplus work—can result immediately in extra production. A work group that is asked to work an additional four hours at the end of an eight-hour shift will obviously add considerably to the day's production. It will probably not be maximally efficient production, but it will be additional production nonetheless.

The same factors that combine to make work sharing economically unattractive as a method for responding to decreases in demand have made it, through overtime, increasingly advantageous in meeting temporary surges in demand. Fixed per capita costs for additional employees are eliminated, as are some taxes (significantly, UI) that are paid only on base earnings.

The fact that so many employers find the temporary use of overtime more economical and efficient than the hiring of additional workers is indicative of the value of the capability for such instantaneous responses when a firm is on a reduced workweek under an STC plan. If increasing the workweek from 40 to 48 hours is advantageous despite the economic penalties of premium pay and a drop in productivity, certainly an increase from 32 to 40 hours should have all of the same advantages and none of the disadvantages, as has been demonstrated by actual experience with STC in Arizona, California, and Oregon.

Obviously the use of overtime is more attractive where workers are highly skilled, where workers with comparable skills are difficult to recruit, and where training and per capita fringe and administrative costs are high. The same factors also enhance the value of work sharing with STC.

All of these factors seem likely to exist to a higher degree in unionized situations and thus to make STC relatively more attractive to unionized firms. For the union business agent or local officer, the only disadvantage of work sharing with STC is that it will slightly depress the earnings of the group of workers who would not have been laid off. To the extent that this single disadvantage is counterbalanced by the increased leisure, greater security, lower taxes, savings in travel, and other work-related expenses sufficient to make STC palatable to the normally influential group of senior workers who perceive themselves to be immune to layoff, STC is apt to be enthusiastically embraced by local union leaders. Such has already been found to be the case among union leaders who have experienced STC in California.[5]

American labor has seldom been interested in revolutionary rhetoric. Instead of talking about taking property from the capitalists, it has been inclined to develop a property right in the job. "Trade unionism," George Bernard Shaw pointed out, "is not socialism. It is the capitalism of the proletariat."[6] What appears to be a steady erosion of the doctrine of employment-at-will may ultimately extend to nonunion workers the protection against arbitrary discharge now enjoyed by union workers. Coupled with the continuing job security promised by STC, this could conceivably give all workers, union and nonunion, an attachment and right to their jobs heretofore unknown in America.

Unions and Shorter Hours

While labor's contemporary interest in full employment has focused on fiscal and monetary policies to create new jobs, the historic means toward the same goal has been through shorter hours. Indeed, the struggle for shorter hours was the history of the labor movement from the early nineteenth century through the New Deal.

The early-nineteenth-century demands for shorter hours were perhaps a product of the continuing impact of the American Revolution. While the desire of a worker for something shorter than a twelve-hour day may seem to require no justification to the modern reader, demands for shorter hours were clothed in egalitarian rhetoric. Philadelphia carpenters, echoing Jefferson's words in the Declaration of Independence, adopted a resolution in the 1830s that asserted: "All men have a just right, derived from their creator, to have sufficient time each day for the cultivation of their mind and for self-improvement."[7]

In New England, the *Boston Transcript* lent ringing editorial support to the movement for shorter hours: "Let the mechanic's labor be over when he has wrought ten or twelve hours in the long days of summer, and he will be able to return to his family in season, and with sufficient vigour to pass some hours in the instruction of his children or the improvement of his own mind."[8]

The late-nineteenth-century fight for the eight-hour day was directed by the Federation of Organized Trades and Labor Unions (the forerunner of the AFL). Its principal argument was not the leisure and self-improvement rationale of the earlier era but rather the creation of jobs for the unemployed. And, as has often been the case in labor's advocacy of social legislation, there was a congruence of altruism and enlightened self-interest. A larger reservoir of unemployed workers represented an increasing threat to the bargaining power of organized workers. Employers, then as now, do not respond well to demands for higher wages, shorter hours, and better working conditions when there are willing hands eager to replace incumbent workers.

New Deal legislation put a dramatic halt to what had been a continuing campaign for shorter hours, and not since the New Deal have there been any ambitious labor campaigns for workweek reductions. The Fair Labor Standards Act made uniform the 40-hour week that had already been established as the norm by the National Recovery Administration. Through collective bargaining (encouraged and strengthened by the Wagner Act), unions made but few efforts to reduce the 40-hour week but instead translated it into the eight-hour day, with negotiated premium pay requirements for any daily hours over eight and for work performed on weekends, holidays, or outside scheduled working hours.

The combined effect of UI and the Fair Labor Standards Act (FLSA) in changing American industry's method of dealing with reduced demands for labor from work sharing to layoff is dramatically revealed in statistics from the 1938 and 1949 recessions. Before the impact of UI and the FLSA, the employer response to recession was a traditional workweek reduction—33.1 hours compared with 42.1 in February 1937—a drop of 23.1 percent. In the February 1949 recession, however, layoffs with UI were used instead of work sharing, and there was an insignificant decline in the workweek—40.5 hours in February 1948 compared to 39.6 in February 1949—a drop of only 2.2 percent. In no recession since that time has there been any appreciable reduction in hours.[9]

Work Sharing, STC, and the AFL-CIO

Despite oft-stated misgivings about work sharing, the AFL-CIO gave its conditional support to STC as an alternative to layoff in an August 1981 statement that "on balance short-time compensation with appropriate safeguards is a worthwhile approach."[10] The action was due in part to favorable reports from California unions participating in the program and in part to the existence of work sharing without STC in many unionized firms.

A Bureau of Labor Statistics survey identified work-sharing provisions in 24 percent of the major collective bargaining agreements in 1980.[11] Thirty-five percent of the workers in the study were affected by such clauses, found mostly in manufacturing, which accounted for almost one-third of the contracts, covering more than one-half the manufacturing workers. Similar high percentages existed in communications (74 percent of the contracts covering 72 percent of the workers).

Contracts with provisions providing for reductions in hours rather than layoff frequently have limitations on the depth and/or length of the reduction. In the steel and telephone industries reductions are limited to 20 percent (32-hour weeks). A United Auto Workers agreement with Twin Disk permits an eight-hour reduction for up to four weeks with layoffs to follow.[12] It is significant that while the Amalgamated Clothing Workers agreement with Xerox limits the reduction to a four-week period, a Xerox plant in California put in place, with union approval, a longer reduction subsidized by STC.

Exceptions to contract rules on both seniority and work sharing go in both directions but are seldom reported except anecdotally. When the Iron Workers local for Northwest Pennsylvania found many of its members unemployed for so long that they were losing coverage under the union's health and welfare plan, they negotiated side-bar agreements in conflict with contract language in an effort to deal with the problem. One contractor was permitted to work two shifts of six hours daily—6:00 A.M. to noon and noon to 6:00 P.M.—thus spreading employment and enabling him to complete a construction project in a more timely fashion. Another construction site was placed on a compressed work schedule of four tens—four days of ten hours, each worked by one crew alternating with another 40-hour, four-day schedule for a second crew.[13]

The business agent was inspired to undertake this unconventional approach (which included waiving a traditional and strongly held union commitment to premium pay for all work over eight hours and on weekends) by stories and memories of how the union had spread the work in the early 1930s.

California business agents and international representatives reported that "with respect to work sharing, 47.3% had used short time *without* unemployment compensation benefits prior to the adoption of UI."[14] The history, then, is one of eclectic response to temporary employment problems. While layoff was extremely common as a response, work sharing had also been experienced by nearly half of the unions.

The prevalence of work sharing without STC has led some to speculate that STC will provide a windfall, at great expense to the trust funds, to those already dividing the hours. But since the adoption of STC requires an employer initiative and will ultimately be reflected in experience-rated (or -surcharged) taxes, it seems safe to assume (and California experience confirms) that undue drains on the trust funds will not occur.

Despite some use of work sharing without STC, experience in California has demonstrated that STC sharply increases its use. Some employers hesitate to implement work sharing without STC because of worker resistance, as reported by Rosow and Zager,[15] and uncertainties as to the outcome of the arbitration of grievances that might arise if they unilaterally reduce the workweek in the face of a seniority layoff clause.[16] Both union policies and industrial relations will be stabilized as STC becomes a known norm for labor and management. As John Zalusky, the AFL-CIO economist who is the labor movement's principal spokesperson on STC, says: "The presently unattractive work-sharing method of adjusting to slack work appearing in so many agreements would become a viable and attractive alternative to layoffs [with STC]."[17] Contrary to the expectation that STC will undermine the concept of seniority, Zalusky suggests that it may provide new opportunities for creative applications of seniority, just as supplemental unemployment benefits (SUB), vacation, and other fringe entitlements are sometimes awarded or scheduled on the basis of seniority.

The AFL-CIO conditional support is cautious and makes clear that the organization feels that STC "does not address the basic need to create full employment,"[18] is not a substitute for the continuing interest in reducing hours by legislation and negotiation, and indeed, merely "masks the underlying problems by simply redefining unemployment numbers."[19]

AFL-CIO Criteria for STC Laws

The AFL-CIO has also spelled out in detail the conditions under which it will endorse specific STC laws.[20] These include:

1. **Adequate financing of unemployment insurance trust funds.** Certainly STC benefits for a minority of workers should not come at the expense of workers who will continue to rely on regular UI. The proportion of wages subject to UI tax has declined severely since the adoption of the law. For example, in Pennsylvania the entire payroll was subject to tax in 1938. The concept of a ceiling on taxable wages was introduced in 1940, and at that time the taxable ceiling of $3,000 covered virtually all hourly wage earners. The $7,000 maximum of 1983 would have to be more than tripled to provide equivalent income.

2. **Approval of the union where there is one.** This is, of course, inherent in the duty to bargain with the certified exclusive agent. AFL-CIO is further concerned with the principle of voluntarism, and even though it has not demanded that the law require approval procedures for unorganized workers, it notes that such requirements are built into the German and Canadian laws.

A requirement for the approval of STC by affected workers might be an important and effective protection for UI as a right. In most states today, an employer may with impunity deprive workers of UI benefits by threatening to permanently discharge any worker who collects UI during layoff. Such a contradiction between a legal and a practical right to UI has led the NLRB to hold that when such punishment of an individual worker has a chilling effect on the collective rights of a work force it constitutes an unfair labor practice.

3. **A wage replacement level of at least two-thirds of the lost pay and workweek reductions limited to 40 percent.** The labor movement has long been concerned that the proportion of wages returned to the unemployed worker has declined in the years since UI was instituted (in Pennsylvania from 48 percent in 1938 to 42 percent in 1983). This is a significantly lower rate of replacement than in most other industrialized nations (in Canada and Germany the replacement is approximately two-thirds) and hits hardest those higher-earning workers who make the largest sacrifice when they opt for STC for themselves rather than layoff for their juniors.

The two-day workweek reduction limit under STC is a practical limit, since with a greater reduction workers would be eligible for benefits under regular UI partials.

The purpose of partials is to create incentive for work, even if it is less than full time. The same principle might be applied in resolving the concern over loss of income which volunteering for STC entails. If STC is as important a public policy as encouraging part-time work through payment of partial UI (some states grant full benefits while permitting some minimal earnigs), several possibilities suggest themselves:

- a. Higher replacement rates for higher-paid senior workers (as one Oregon employer suggested).
- b. No moonlighting restrictions (as Arizona wisely agrees).
- c. Subsidy to the employer for part of the cost of maintaining pensions and health insurance, as though full workweeks were being worked (as Germany does).
- d. Elimination of a waiting week. The waiting week decreases the replacement rate and is an especially onerous penalty for workers who have accepted a work/income reduction for the benefit of others.
- e. Eliminate the charging of STC benefits drawn by workers (who 80 percent of the time would not have been laid off) against their total entitlement (as Canada does). If STC leads to ultimate closure, as it may in some cases, they will need and deserve their full-term benefits.

4. **Retention of fringe benefits.** The continuation of fringes together with the partial wage replacement through STC were the critical considerations for workers and unions revealed in the 1980-81 Arizona survey.[21]

Health insurance will normally represent the highest cost for employers, and consideration might be given to the subsidization of such additional costs.

Since the payment of most pension benefits is based on earnings, these will not normally represent any additional costs to employers during periods of STC. AFL-CIO proposes that workers who volunteer for STC not have their pensions penalized. Many pension benefits are determined by a worker's average salary during the last few years of employment, and this could represent a knotty problem for some workers. As a result of participating in an STC plan, a worker nearing retirement could have his or her pension reduced throughout the retirement years and thus lose far more than the wages lost during the period on STC.

One solution for such problems may lie in the area of flexibility and voluntarism. In at least one case, a union business agent arranged for a few workers nearing retirement to be excluded from the STC work reduction group and to be fully employed on maintenance or other work so that their pension benefits were not adversely affected.

5. **Prohibition of discrimination against recently hired workers, especially minorities and women.** "[This] objective," says Zalusky, "is the most fundamental of all. We want to insure that women, young

people, and minorities are able to hold on to the jobs that they have struggled to obtain."[22]

The AFL-CIO's emphasis on this goal is consistent with the aims of the earliest American advocates of STC as a means of protecting affirmative action gains.

AFL-CIO spokesman Zalusky asserts that it cannot be

> simply assumed that STC will improve opportunities for minorities, youth, women . . . [because] in most real world of work situations the newly hired worker is easily recruited and replaced. The employer's investment in the recently hired worker's training and on-the-job experience is relatively low and the replacement cost is minimal. On the other hand, the senior workers are likely to be more highly skilled, have greater employment opportunities within the locality and industry, and are much more costly to locate, recruit and train.
>
> [The labor movement wants] strings attached to insure that women and minorities are protected. [Suggested strings include requiring] that the employer not have laid off any employees within six months prior to application for certification [or] that the organization's progress (toward its affirmative action goals) has not been adversely affected by prior layoffs.[23]

Union Experiences with STC in California

One of the most outspoken union advocates of STC in California was an IAM business agent who even went out of state to preach its virtues. But he never failed to include a warning that STC was only a temporary palliative and that a shorter workweek, with greater penalties for overtime, was needed to create as well as protect jobs.[24]

As more workers experience the STC program, the value of increased leisure may become more apparent. The virtue and value of leisure was one of the unanticipated benefits of STC for many unionists. One representative reported that even though it was not an anticipated advantage when STC was initially considered, leisure time was now considered by the workers to be of major importance. Another business agent pointed to the creativity with which the union and employer could use STC to permit extra-long weekends by scheduling three days in one week and five in the next. For the first time, four-day fishing trips became possible for his members without using up regular vacation days.

Another union representative was most articulate in describing what he believed were the real leisure-time benefits. For the first time, he explained, a father could go to school to pick up the kids on Friday and see what that part of their life was like. Husbands could spend more time at home making repairs and take more time to shop, thereby becoming more intelligent consumers. In his experience, many of these men used to spend their weekends, their only time off, drinking with friends and relatives. But on STC days off this doesn't happen because neighbors or relatives aren't free on the same day. His members, he believed, were more relaxed workers who made fewer mistakes. In short, the impacts on the psychological state of the worker and on the worker's family were very positive.

One business agent perceived a leisure-related benefit that also affected the health and welfare fund. Workers were covered for dental benefits, but seldom took advantage of the preventive care available because their weekends off were also the dentists' time off. They waited until problems became so painful that they were forced to miss a workday. This cost the fund money, the worker wages, and the employer production. Scheduled days off resulted in better dental habits, long-range savings on major dental bills, and less absenteeism on scheduled workdays.

Layoffs threaten the union not only with short-range losses in dues income but with the demoralizing and disunifying effects on a work force in which some have lost their jobs and the rest feel threatened. Instead of focusing its energies on the causes of this affliction to the collective it represents, the union must devote its energies and resources to the impact on individuals. Its members are not all together at the workplace; they are harder to find, organize, and defend.

Moreover, as several business agents were quick to point out, there is seldom a "pure seniority" contract provision with respect to layoff. The usual clause provides for some interaction between seniority and ability and leads to constant tension within the union and potential fair-representation problems in handling grievances. One agent, experienced with mass layoffs, spoke with emotion about the problems of union officials with provisions for "last-hired, first-fired and all that crap." He maintained that grievances in such situations were never resolved to everyone's satisfaction and that the problems of conflicting interests were nearly impossible to resolve. He said, "The company doesn't care if Joe or Sam does it, as long as he's qualified and capable. They don't want to spend time and money on grievances and arbitration. Joe and Sam each care—each want the work—and the union is worried and torn."

Business agents reported a high degree of satisfaction with STC assistance in maintaining union membership and strength. This was the key advantage to STC for at least one of them, since, in his view, "a reduction in numbers means a reduction in strength."

Union leaders, of course, are also concerned about the political problem of membership support. Turnover in membership creates problems of loyalty, unity, and membership support. One representative who had recently been defeated for reelection to office said that if all voters in the local were as supportive of him as the workers in the shops using STC at his suggestion, he would have won. The program had created for him a constituency.

Another representative pointed out that the program offered unions a chance to tell their members "what they've done for them lately." Simply negotiating and enforcing the collective bargaining agreement during good times but appearing impotent and unhelpful at the moments of crisis in workers' lives was not good unionism in his eyes. From his point of view, the program allowed business agents to demonstrate their expertise and initiative to the membership and was therefore very useful politically.

STC is seen by most union officials interviewed as a small but significant effort toward answering the question posed in the old labor song, "How can I work when there's no work to do?"

Julius Uehlein, president of the Pennsylvania AFL-CIO, who had observed STC at work on his visits to European unions, said: "You can't just sit back and hope unemployment will go away. You have to try new approaches."

NOTES

1. Statement by the AFL-CIO Executive Council, 5 August 1981.

2. Ibid.

3. Sumner H. Slichter, *Union Policies and Industrial Management* (Washington, D.C.: The Brookings Institution, 1941), chapter on "Control of Layoffs—Union Policies."

4. Article in *United Automobile Worker*, 16 February 1938, p. 6, cited in Carl Gersuny, "Origins of Seniority Provisions in Collective Bargaining," *Labor Law Journal*, August 1982, p. 523.

5. Martin J. Morand and Donald S. McPherson, "Union Leader Responses to California's Work Sharing Unemployment Insurance Program," paper presented at the First National Conference on Work Sharing Unemployment Insurance, San Francisco, 15 May 1981 (*Daily Labor Report*, 28 May 1981, pp. C1, D10).

6. George Bernard Shaw, *The Intelligent Woman's Guide to Socialism and Capitalism* (New York: Brentano, 1928).

7. George Brooks, "Historical Background," in *The Shorter Work Week*, a collection of papers delivered at the Conference on Shorter Hours of Work, sponsored by the AFL-CIO (Washington, D.C.: Public Affairs Press, 1957).

8. Ibid., p. 8.

9. Ronald W. Schatz, *American Electrical Workers: Work, Struggle, Aspirations, 1930-1950* (Pittsburgh, Pa.: University of Pittsburgh, 1977), p. 147.

10. Statement by the AFL-CIO Executive Council, 5 August 1981.

11. U.S. Department of Labor, Bureau of Labor Statistics, *Major Collective Bargaining Agreements, 1980* (Washington, D.C.: Bureau of Labor Statistics, 1980).

12. John Zalusky, "Short-Time Compensation," remarks to the Interstate Conference of Employment Security Agencies, Inc., 23 September 1982.

13. "Ironworkers Turn to Work Share," *Morning News* (Erie, Pa.), 15 June 1962, front page.

14. Morand and McPherson, "Union Leader Responses."

15. Jerome M. Rosow and Robert Zager, "Punch Out the Time Clocks," *Harvard Business Review* 62 (March-April 1983): 26.

16. Michael J. Haberberger, "The Arbitration of Worksharing Disputes: Developments and Implications," *Labor Arbitration Information System: Perspectives*, 10 (May 1983).

17. Zalusky, "Short-Time Compensation."

18. Ibid.

19. Ibid.

20. Ibid.

21. Zeena Weber, Joseph T. Sloane, and Robert D. St. Louis, *Shared-Work Unemployment Compensation: Arizona Survey Results* (Phoenix, Ariz.: Arizona Department of Employment Security Unemployment Insurance Administration, 1981).

22. Zalusky, "Short-Time Compensation."

23. Ibid.

24. All of the following reports on California union experiences with short-time compensation are from Morand and McPherson, "Union Leader Responses."

II
SHORT-TIME COMPENSATION: AT HOME AND ABROAD

4.
The Pioneers: STC in the Federal Republic of Germany

Harry Meisel

In order for the specifics of the short-time compensation (STC) program of the Federal Republic of Germany to make sense to an American audience, a brief background of the political, economic, and societal environment in which it functions may be helpful. STC (in Germany it is "Kurzbarbeitergeld-Kug," but I shall use the English acronym STC) is one of many employment programs administered by the Federal Employment Institution. The concept of STC in Germany is now more than half a century old. It originated in 1927 and during the postwar years has been modified frequently and adapted to current needs. Present procedures were established in 1969 by the Employment Promotion Act.

Unemployment is viewed most seriously in the Federal Republic of Germany. It is seen not simply as an economic problem but as a social, political, and even foreign policy issue as well. This is the major Western country sharing a border with the Warsaw Pact countries, and along that border the rate of unemployment is an immediate and meaningful reflection of the "score" of the continuing East-West political and economic competition.

During periods of increasing unemployment, German political leaders are particularly mindful of the fact that the overthrow of our last democratic government before Hitler came to power in 1933 was directly attributable to heavy unemployment with which the government was unable to cope.

STC and other unemployment benefits should be viewed in the context of our general tripartite employment policy, sometimes called the social contract. Government, labor, and management are partners, and my own office reflects and illustrates that partnership. The Federal Employment Institution of Baden-Wurttemberg is administered by a 27-person board, with members drawn equally from management, labor, and government (federal, state, and local). As president, I am elected

53

by the board and appointed by the President of the Federal Republic upon recommendation of both the federal and state governments.

STC enjoys the unanimous support of all three groups on the board because of its demonstrable contribution toward maintaining stability of employment. Such stability contributes to the general stability of labor-management relations as well as to a broader societal stability.

STC is part of our unemployment compensation system, which is financed by a contribution of 4.6 percent of wages. The contribution is divided equally between worker and employer, and if an emergency depletes the fund, the federal government would provide grants and/ or loans. Even though individual states enjoy great autonomy in most matters and most federal laws are administered by state or local governments, the entire field of work-force and labor-market activities is not only regulated by federal legislation but is also federally administered so that the Federal Employment Institution of Baden-Wurttemberg has an essentially identical counterpart in each of the other states. The 4.6 percent payroll contribution finances the entire operation, providing funds not only for unemployment compensation, unemployment assistance, and STC but also for vocational guidance, training, and placement. In general, the unemployment compensation system is designed to replace over two-thirds of a worker's net loss of income due to layoff. And STC is designed to provide the same percentage offset against loss. The budget in 1982 was approximately three billion deutsche mark ($600,000,000).

The Employment Promotion Act of June 1969 spells out the somewhat flexible guidelines for the establishment and administration of short-time compensation plans:

> Short-time compensation shall be granted to employees in the event of a temporary shortage of work in any establishment where at least one worker is regularly employed, if it can be assumed that the payment of such compensation will enable the employees concerned to retain their jobs and the establishment to retain experienced labor. If there is a substantial labor shortage, such compensation shall not normally be paid when the employment market situation mandates the placement of the employees concerned in other employment that they can reasonably be expected to accept.

The act identifies the three-fold goal of STC:

1. To assist the employer in preserving the human capital represented by labor and in avoiding the costs associated with layoff and termination as well as in preventing the expenses of labor recruitment and training.

2. To enable workers to protect their property interest in their jobs

and to reduce the income losses which would result from total unemployment.

3. To provide economic stability by sustaining demand.

Establishing an STC Program

The most common but not exclusive procedure for the adoption of an STC plan is for the employer to make application to the local employment office. The act provides that such application must include:

1. The starting and probable ending dates of reduced hours.
2. The regular hours of work.
3. Extent of the proposed reduction in hours.
4. How full-time and reduced work hours will be mixed during the period of reduced hours.
5. The total number of employees.
6. The number of affected employees.
7. Reasons for the reduction. This includes details related to the business necessity behind the reduction (flow of orders, shortage of raw materials, and production scheduling) as well as information on measures taken to attempt to avoid the reduction (building inventory, stockpiling of raw materials, assigning workers to maintenance and housekeeping duties, and scheduling vacation and other paid leave). The view of the works council (which may, but in practice seldom does, initiate the application on its own) must be reported.

Each firm's works council consists of employee representatives and is the workplace application of the German codetermination concept. The works council representatives are generally members of the trade union, and their consent to an STC application binds all the involved workers. In the absence of a works council, all affected workers must approve the application.

The works councils are major factors in preventing abuse of STC. They check to be sure that a reduction in hours is inevitable and are usually well informed about the business and the steps that have been or should be taken to avoid the necessity for either layoff or STC.

After an application has been made, the local employment office must review and approve the application before its adoption. The application will be rejected where the firm does not regularly employ at least one worker or in cases where regular work or regular hours of work are not the norm. Most employees in the entertainment industry, for example, are not covered by STC because of this requirement.

In principle, STC is paid only where the shortage of work is temporary because otherwise the goal is to place the workers in alternate suitable employment—particularly if there is a labor shortage in other enterprises. But even in such cases, overwhelming public policy concerns may lead to ignoring technicalities in the interest of practicality. For example: a small company near the border of East Germany employs several skilled workers, but its work force is predominantly unskilled, part time, and female. Other jobs could readily be found for the skilled workers, but their absence would leave the bulk of the work force unemployed in a highly sensitive geographic area. So the skilled work force is not dissipated by placement elsewhere: they are granted STC benefits, and the jobs of the unskilled are thus preserved.

Certain minimum standards are imposed to discourage a firm's use of STC in situations that are of very short duration or that involve a very slight reduction in hours for only a few workers. During the first four-week period following the date of application for STC, at least one-third of the firm's employees must experience more than a 10 percent reduction in working hours, and the total working hours of the entire work force, whether affected by short time or not, must be reduced by more than 3 percent.

Once the firm or organization's STC application is approved by the local employment office (over 95 percent of all applications are approved), the individual affected worker gains an entitlement to benefits. The worker's eligibility is conditioned on continued employment by the firm but for fewer hours and less pay. Cessation of work—whether to enter a vocational training program that provides a maintenance allowance or to take paid leave such as vacation or sick leave—will render the worker ineligible for STC benefits. STC payments are reduced by 50 percent of net outside income (after discounting for tax reductions, social insurance payments, and other expenses of going to work). After such adjustments the worker is entitled to 68 percent of the *net* wages or salary that he or she would otherwise have been paid. As with regular unemployment compensation, family and children's allowances are added.

STC is payable for up to six months, but this period may be extended to 12 months in cases of high or prolonged unemployment in certain industries or areas. If the entire economy is distressed, the benefit period may be extended to 24 months. Once benefits have been exhausted, three months must elapse before eligibility can be reestablished. On December 20, 1982, legislation was adopted for the years 1983 and 1984, providing for an extension to 36 months of benefit eligibility for the steel industry.

The procedure for making STC payments is simple. The employer pays the worker and is reimbursed by the local employment office. Since the employer is required to maintain health and pension benefits as if the worker were fully employed, the employer is also reimbursed for

50 percent of the additional health insurance payments and 75 percent of the additional pension fund contributions it makes because of this requirement.

Administering Short-Time Compensation

The administration of STC is both simpler and less costly than the administration of regular unemployment compensation benefits. Under the regular system, the individual worker must visit the local unemployment office and have individual benefit checks written and transmitted to him or her. The individual must be directed and frequently helped in filling out the information forms, which must be verified subsequently against employment records; must be assisted in searching for alternate employment, a process that may involve skills assessment as well as vocational counseling and training; must have his or her application for benefits evaluated for eligibility and the appropriate benefit level determined; and must make periodic reports to, and keep in contact with, employment office personnel. This extensive use of professional personnel on an individual basis is in sharp contrast to the minimal or nonexistent need for such professional personnel services in the STC application, approval, and payment system. The largest part of the paperwork is performed by the employer, who transmits the application and payroll information and calculates and pays the benefits. Nor is this an undue burden on the employer because it must provide essentially the same data to the tax collector and Social Security system. Indeed, an identical computer program printout is frequently used for STC and other reporting or accounting purposes.

As with any other program, it is, of course, possible for STC to be abused. Investigations by local employment offices, which have a right to examine the records and visit the premises of participating firms, have uncovered only rare, willful misuses of STC. There is a built-in restraint against deliberate fraud or even negligence, since the employer is financially liable for any benefits improperly paid.

There are two additional natural controls against abuse. The works councils and the trade unions continually press for full employment and will normally resist unnecessary reductions in hours. The employer utilizing STC experiences a slightly higher direct unit labor cost because of the unreimbursed portions of health insurance and pension contributions and thus could not afford to unduly prolong STC when a permanent reduction in force was mandated by business necessity. Workers on STC rarely seek part-time jobs or vocational training, and few employers would want such temporary, part-time workers.

It is possible that in certain isolated instances the use of STC has delayed a necessary reduction of the labor force. This may have been partly because it is the nature of entrepreneurs to be optimistic and partly because of the German postwar experience of sustained economic growth. For whatever reasons, the use of STC in these few cases may well not have been the best or wisest in terms of strict efficiency. Due to the costs of unemployment to individuals, families, and society as a whole, it is the policy of federal employment offices to lean in the direction of avoiding unemployment to the maximum extent possible.

While public-sector employers and employees are eligible to participate in STC plans, German employment offices have had little experience in administering STC for public employees, since in the public sector it is unusual to have either a layoff or a reduction in hours. The general public policy of avoiding unemployment at almost any cost obviously has its greatest impact on public employers.

One might reasonably assume that a company in danger of bankruptcy would not be a good candidate for STC, since it would seem to be unable to meet the legal requirement that STC be used only where it will lead to continuing employment for at least most of the workers. In at least one instance, however, in a firm where the court had already appointed a trustee in bankruptcy, it was possible through the use of STC to keep the work force together until a new owner could be found to keep the firm in operation.

The laws and administrative procedures governing STC in the Federal Republic of Germany are sufficiently flexible to permit an employer to reduce its work force in one part and place another part on STC (there are, however, restrictions on a reduction in the work force contained in other sections of the labor law). Even the hiring of necessary workers can occur during STC if such hirings contribute to the overall efficiency and viability of the firm.

Despite the flexibility permitted by the law in regard to hiring and discharge during STC, there may be provisions in a collective bargaining agreement restricting the employer's freedom of action in these areas.

Collective bargaining agreements also occasionally cover the conditions under which STC may be applied. Sometimes the union contract will establish such liberal supplemental unemployment benefits that workers and unions will have no incentive to use STC. Other contracts provide that the payment of supplemental benefits be triggered by the introduction of STC.

Short-Time Compensation in Baden-Wurttemberg

The state of Baden-Wurttemberg, located in the southwest of the Federal Republic, is large and highly industrialized. Twenty-four district and 80 local employment offices, staffed by some 6,000 employees, are administered by my office.

The state of Baden-Wurttemberg has a higher usage of STC than the Federal Republic of Germany as a whole. This may be in part due to the concentration and variety of industry (automobiles, mechanical and electrical engineering, textiles, and clothing) in the state. Smaller employers have sometimes been hesitant to use the program because of social pressures. There was a sense, particularly in the recession of the mid-1970s, that adopting STC would diminish the reputation of the firm and might even hurt credit ratings, particularly in smaller communities where everyone is presumed to know everyone else's business. Applying for STC was thought to imply that the business was not being managed very carefully. Over time, both management and workers have become accustomed to STC, and it is now used with less hesitation.

The higher utilization of STC in Baden-Wurttemberg may also be partly due to the very positive attitude toward the program by the staff at local employment offices and the fact that a positive experience with it in one firm may lead to its adoption in neighboring firms.

A comparison between the state of Baden-Wurttemberg and the Federal Republic, as of the first quarter of 1983, is shown in table 4.1: It is my opinion that without STC the unemployment rate in Baden-Wurttemberg would be significantly higher.

The calculation of STC benefits is based on the earnings of the individual for the 20 working days immediately preceding the adoption of an STC plan. These earnings are expressed in terms of an hourly wage established by the collective bargaining agreement or by practice. Where individuals are paid a weekly, biweekly, or monthly salary, the salary is expressed as an hourly rate. The sole exception to this procedure is the case of pieceworkers, whose earnings may fluctuate wildly in the short run. The hourly rate for them is determined by dividing total earnings for the prior 60 working days by total hours worked.

Once the gross hourly rate has been established, a net hourly rate is taken from a standard table that takes into account graduated income

Table 4.1. STC in Baden-Wurttemberg and in the Federal Republic
as of the First Quarter of 1983

	Federal Republic	Baden-Wurttemberg
Number Unemployed	2,469,803	249,065
Unemployment Rate	10.1%	6.5%
Number on STC	1,120,936	192,736
STC Percent of Unemployed	45%	77%

taxes and Social Security contribution. The benefit rate is 68% of this net hourly rate. (In Germany wages or salaries in excess of $11 per hour or $440 per week do not yield higher unemployment benefits.) STC benefits are themselves not taxed.

In order to judge accurately the administrative costs of STC, eight local employment offices in Baden-Wurttemberg kept detailed personnel time records in 1981. In a typical medium-sized office during that year, 46.8 million deutsche mark were paid in unemployment compensation, and the task of making these payments required the service of the full-time equivalent of 17.6 employees. In the same office the payment of 9.1 million deutsche mark in STC benefits was handled by 2.4 employees. Thus the STC program is much more cost efficient than the unemployment compensation program. But this is only a fraction of the cost differential because it deals only with the cost of paying benefits. There are significant additional (and more expensive) administrative costs in regard to laid-off workers, who must come to the office for counseling, assistance, and possibly training in connection with a new job placement.

Nor are these greater efficiencies of administration of STC limited to the government agency. Employers using computers find it as easy or easier—and more economical—to calculate and report STC benefits as to respond to inquiries with regard to each of those individual employees who might have been terminated if not for STC.

5.
The California Experiment

John Lammers
Timothy Lockwood

Since the time of John Maynard Keynes and the New Deal, the U.S. government has assumed responsibility for a goal of full employment. The government has used three essential weapons to battle unemployment; two are well ensconced, and a third is now burgeoning. The first and most traditional is indirect: incentives to business are expected to create more jobs by increasing total economic activity. The second solution involves the state through direct job creation by expanding public-sector employment and job-training programs. The third approach, and the one receiving increasing attention lately, consists of work-spreading policies that redistribute work opportunities on a reduced basis throughout the labor force. California's Shared Work Unemployment Insurance Program (SWUI), enacted in 1978, is one example of this third approach to employment policy.*

Ever since the initial flurry of legislative activity in 1978 that brought the California program into existence, it has been continuously evolving. The tensions, adjustments, and growth documented below can provide an example of how one state developed an STC program.

---

* The terms "shared work," "work sharing," and "short time," coupled with "unemployment compensation" or "unemployment insurance," have all been used in different times and in different places to designate a planned program of employer-managed reductions in employee workweeks (and wages), which is accompanied by state payments of prorated unemployment insurance benefits to the workers in partial compensation for their lost income. Shared Work Unemployment Insurance (SWUI)—and sometimes Work Sharing Unemployment Insurance (WSUI)—are the titles given to the California program but the acronym STC will be used in this chapter for both the California program and similar programs elsewhere.

Shared Work Legislative History and Early Administration

Employer voluntarism and flexibility are the key features of the original California legislation. The legislation was designed to include only the modifications to the existing Unemployment Insurance (UI) Code deemed absolutely necessary to create an STC program. The legislation added only two sections (two pages) to the California UI Code, and was enacted as "urgency legislation." Once introduced, California Senate Bill 1471, authored by Senator Bill Greene, passed through the legislative process in less than 36 hours.

The specter of immediate and permanent layoffs in the public sector, primarily due to the impending passage of Proposition 13 property tax reforms, was largely responsible for the swift adoption of STC. A purported alternative to those same layoffs, the STC program was originally intended to be a short-term, experimental emergency program to ease workers out of public-sector employment. Public employees on STC, it was reasoned, would have additional time to search for work elsewhere before their threatened jobs disappeared altogether in post–Proposition 13 budget cuts.

Two events altered the purpose of the California program. First, immediate public-sector layoffs never materialized. The reasons were many. By utilizing a budget surplus accumulated over several years, sporadic and temporary hiring freezes, attrition, and the tax revenue from a relatively stable state economy, the public sector avoided the necessity of either layoffs or STC. Second, the clientele of STC broadened when labor unions and business, both presumed skeptical of the concept for differing reasons, began to show up on the STC enrollment lists of the California Employment Development Department (EDD). The short recession in spring 1980 and the more persistent recession of 1981–1983 have encouraged California's private sector to investigate STC in increasing numbers.

Administrative rules were worked out after the legislation was passed. By August 1978 (one month after legislation), rules for establishing the operation of the program had been developed and guidelines were sent out to California's 150 Unemployment Insurance Field Offices. Due to the rapidity with which STC was implemented, many administrative procedures had to be worked out on an ongoing basis. At the First National Conference on Work Sharing Unemployment Insurance in May 1981, one administrator reminisced:

> We were advised that we had two weeks to put together a program, develop policy, write procedures, design forms, train over 2,000 people in the 140 field offices in a state 1,000 miles long. We approved the first plan in two weeks and issued the first work-sharing payment in the third week . . . we

were not able to give enough attention to the minor details and exceptions that you could anticipate in a normal UI program because we didn't know what they were.

A great deal of discretion was left to the administrators of the program. STC began as a wrinkle in the regular UI system but has grown in scope and achieved administrative autonomy.

The program was set up to operate as follows. Upon determining the need for a labor-force reduction, a California employer writes or calls the EDD central office in Sacramento to request information and application forms. The employer application must specify the total number of employees in the company or agency work force, the total number proposed to share work, and the employer's name, business address, and UI identification number. In addition, any employer whose workers are formally represented by a union or employee association is required to obtain the signature of the appropriate union representative on the application.

The law itself includes a clause stipulating that employees under a union agreement must have the union business agent or representative approve in writing the use of the program. This was meant to protect the integrity of union contracts, a concern of the national AFL-CIO. (The legislation also included a surcharge tax schedule to protect the program from employers who might abuse it, a concern of the business community and unemployment insurance economists.)

Finally, the application must list the names and Social Security numbers of all employees scheduled to share reduced hours and the scheduled weekly reduction planned for each worker. This information constitutes a "work-sharing plan" and is sent to the central office in Sacramento for approval.

Upon receipt of a proposed work-sharing plan, the central office determines whether the plan meets the following criteria. At least two persons must be included in the plan. Additionally, these two persons must represent at least 10 percent of either the employer's total work force or 10 percent of a work unit designated for STC by the employer. Finally, the work reduction experienced by the workers must be at least a 10 percent reduction in hours and wages (one-half day in a 40-hour week).

If those conditions are met, the employer is sent a supply of forms for individual workers to fill out and present to local unemployment insurance (UI) field offices in order to open an STC claim. After opening a claim, workers are permitted to file weekly claims and collect benefits by mail. While STC information services, policymaking, and plan approval are centralized, the payment of benefits is localized in field offices.

In California employers are charged an annual UI "Contribution" payroll tax of from 0.0 to 3.9 percent on the first $6,000 paid each employee. The rate varies with the solvency of the UI fund and the employer's experience rating, the ratio of contributions paid by the employer to benefits paid to the firm's laid-off employees. In California, the ratio of the employer's reserve balance to the average payroll is used to compute the experience rate. This is a basic feature of the UI system, but it varies slightly from state to state. When the proportion of UI funds to taxable wages falls below 2.5 percent, a higher tax schedule is imposed. Under the California STC program, an additional surcharge ranging from 0.5 to 3.0 percent of adjusted payroll is placed upon certain employers with poor experience ratings. These taxes are assessed yearly but collected quarterly.[1] The complexity of the surcharge formula becomes apparent when one considers that layoffs made several years previously may affect the experience rates under certain circumstances.

The amount of benefits workers may collect under the STC program is simply a prorated portion of the benefits to which they would have been entitled had they been laid off and claimed full UI benefits. As of 1982, California's regular UI weekly benefit ranged from $30 to $136, depending upon a worker's highest quarterly earnings in the past year, dating back from the quarter prior to the time of layoff or workweek reduction. Thus, the benefits awarded to workers under STC plans for a one-day-per-week reduction in work time ranged from $6 to $27.

Workers are presently permitted to draw STC benefits for 20 weeks during any 12-month period. The employer, however, can spread work continuously as long as work reductions are rotated throughout the work force to avoid exhausting any individual worker's annual eligibility. Workers who exhaust their STC eligibility remain eligible to receive the balance of their regular UI entitlement minus the number of days of STC benefits. If, for example, an employer used the program for 20 weeks, reducing the workweek by one day, and subsequently laid off an employee, that worker would retain 22 weeks of regular UI eligibility (26 weeks minus 20 days).

These are the features of the program that became available to California employers and employees in the late summer of 1978. Even though a few adjustments (which will be examined later in this chapter) have been made to the program since that time, the basic framework remains.

Growth of the STC Program, 1978–1982

Although EDD managed to install administrative guidelines quickly, the program was used by only a few employers in 1978. Because the new program encountered resistance from a few conservative administrators inside EDD and protests from the California Chamber of Commerce against the payment of UI benefits to partially employed people (and was expected to be opposed by union leaders), EDD's initial advertising campaigns in 1978 and 1979 were rather understated. The department took a guarded position on STC, awaiting the outcome of the state's evaluation.

Despite the failure to energetically publicize the program at its inception, it has grown steadily, as shown by table 5.1.

Average monthly employer enrollments jumped from only 3 in 1978 to 40 in 1979. This figure increased again in 1980 to an average of 75 employers per month. In 1981, the figure appeared to level off at an average of 65. However, in 1982 the average monthly number of employers signing up to use STC increased 331 percent over 1981 to an average of 213 employers per month. The growth in the number of *employees* enrolled in the program parallels these figures: a jump from 1978 to 1979 that leveled off through 1981 and a massive expansion in 1982.

What is the explanation for this pattern of growth? Any large undertaking, from social programs to commercial ventures to military maneuvers, requires time to "come up to capacity," and this may have been particularly true of STC, since implementation occurred so rapidly without time for adequate planning. Another possible explanation may be the fact that the use of STC so closely paralleled California's general unemployment trends. The number of job losers in California jumped from 400,000 in January 1981 to over 600,000 in January of 1982 and surpassed 800,000 before January 1983.[2] Similarly, initial claims for unemployment insurance (i.e., regular UI available to job losers without work), after appearing to

Table 5.1. STC Program Growth. Monthly Averages of Employers and Employees Enrolled in STC Each Year, 1978–1982

| Year | Employees Enrolled | | Employers Enrolled | |
	Average Each Month	Yearly Total	Average Each Month	Yearly Total
1978*	106	640	3	16
1979	749	8,245	40	474
1980	2,510	30,122	75	896
1981	3,226	38,731	65	785
1982	8,278	99,332	213	2,567

* Because the program was initiated in July 1978, only six months are counted in that year.

Source: The figures in the table have been tabulated from various issues of EDD, "Work Sharing Activities Monthly Reports."

level off in 1979, increased dramatically in early 1980, appeared to level off in 1981, and then increased steadily again through 1982.

However, there are other reasons that the program grew as it did. One clue is in several dramatic increases observed in worker enrollments from month to month between early 1979 and late 1982 (see fig. 5.1). Why these irregular leaps? An overall increase may be attributed to the declining economy, but some portion of the growth is due to the gradual increase in the size of STC employers. As we shall see, this may in turn be attributed partly to efforts on the part of EDD to publicize the program.

Referring to Figure 5.1, aside from an erratic increase in monthly enrollments in the program from 1979 through 1982, several dramatic slides and peaks may be observed among employee enrollments, notably in June and August 1980, June and October 1981, and October 1982. These increases can be explained partly by comparing the number of employers to the number of employees enrolling in the program each month. It is clear that in midyear 1980, 1981, and 1982, employers with large work forces were enrolled. This feature became firmly established in 1982 (compare employer and employee growth lines in fig. 5.1). Indeed, the proportion of large firms using STC (those employing over 100 workers) did increase from 1980 through 1982.[3] This increase in the use of STC by large California employers partly explains the great increase in employees eligible to receive STC benefits.

The sustained increase in employer enrollments in 1982, however, is as likely due to a new and energetic communications officer within EDD as it is to the recession of 1982. Early publicity in the newspapers had been supportive of STC, and the union and employer opposition, once feared, never materialized. Convinced that there was nothing limiting the program but the awareness of California employers, the new publicist undertook an aggressive campaign to advertise the program. In the ten-month period between February and October 1982, six increases in STC enrollments were recorded.

News releases were sent periodically to every wire service, major newspaper, trade journal, chamber of commerce, trade association, and major television and radio broadcaster in the state. In addition, radio and television announcements were produced and aired. EDD staff members appeared on local and national television programs, and testimonials were solicited from employers who had experience with the program. Certainly, such a campaign to increase public awareness of the program played a part in its sustained high growth in 1982. Concurrently, California's rate of unemployment rose in 1982, introducing double-digit unemployment to Californians by summer. Publicity and need were interwoven.

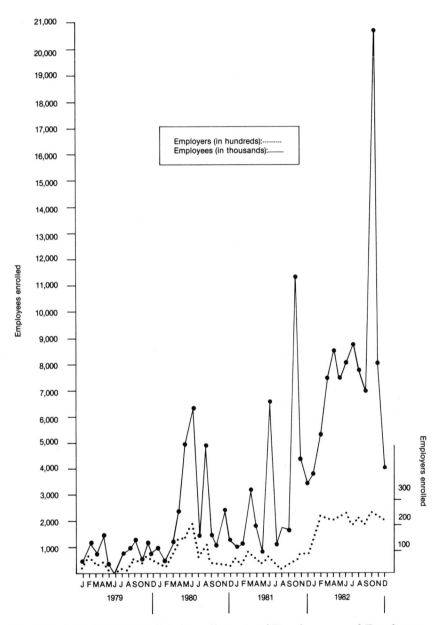

Fig. 5.1. Initial STC Monthly Enrollments of Employees and Employers,
1979–1982
Source: California Employment Development Department, "Work Sharing Activities Monthly
Reports," various issues.

From the point of view of the public information office, the STC program was a winner by 1982, a tested device to fight unemployment. Aggressive promotion of this state service appears to have had a strong legitimizing effect. Within EDD, STC came to be seen as a service of proven merit. As one manager put it: "We *do* have products. I sell my products the same way I would sell soap or cereal. I sell services. Some products are better than others. This is one of our extra special good products."

By the end of 1982, the STC program was no longer an experiment with an uncertain future. At the close of the fourth quarter of 1982, nearly 15 percent (34,192 of 233,100) of all initial UI claims were STC enrollees.[4]

Who Shares the Work: Characteristics of Employers

The general characteristics of employers (referred to as "users") of the STC program include the size of work force, industrial sector, and whether or not the employer is unionized. These employers will be compared with the general California employer profile and contrasted with California industries that utilized layoffs rather than STC.

Before proceeding, it is important to qualify these comparisons. For employers who use the STC program, the primary concern is saving labor costs while retaining valued employees. A number of employment strategies are available to an employer facing pressure to reduce costs. Since no employer has an obligation to retain all employees, layoffs are commonly utilized to prune labor costs. Other strategies allow natural attrition to accomplish reductions. Others reduce costs by adjusting non-labor costs in order to "hoard labor." Still others reduce labor costs by sharing reduced work (either with or without partial UI compensation).

In California, employers can opt to cut hours (unless forbidden by collective bargaining) without offering STC benefits. STC employers, consequently, are those who must make labor adjustments more rapidly than worker attrition would allow, but who do not want to risk the loss of skilled employees through full-time layoffs. Additionally, these are employers who wish to hold their employees but cannot or will not "hoard" them on a full-time basis. While employing them less than full time, they are willing to pay for the partial UI benefits that STC provides, thereby reducing labor costs and maintaining employee attachment simultaneously. A ranking of employee value in terms of layoff strategies might place "hoarded workers" at the top, STC enrollees next, work sharers without UI benefits next, and laid-off workers at the bottom.

What does this mean for comparison of STC and regular UI? It means that, strictly speaking, the two are not merely alternatives to each other, but under similar conditions of economic necessity, they are preferred

Table 5.2. Comparison of Employers by UI Charges within Industry, Size, and Wage Categories[a]

	Noncharge		STC		Regular UI		Total	
Industry	Percent	Number	Percent	Number	Percent	Number	Percent	Number
Total	100.0	286,186	100.0	714	100.0	217,821	100.0	504,721
Construction	6.1	17,584	7.0	50	14.9	32,368	9.9	50,002
Manufacturing	5.2	14,802	45.0	321	11.4	24,826	7.9	39,949
Trade & Services	69.5	193,795	34.2	244	54.0	117,609	62.7	316,648
Other	19.2	55,005	13.9	49	19.7	43,018	19.4	98,122
Number of Employees								
Total	100.0	286,963	100.0	714	100.0	217,837	100.0	505,514
1–25	99.4	285,162	51.1	365	82.3	179,259	91.9	464,786
26–100	0.6	1,745	30.8	220	13.8	30,062	6.3	32,027
Over 100	0.0	56	18.1	129	3.9	8,516	1.7	8,701
Total Wages								
Total	100.0	286,963	100.0	714	100.0	217,837	100.0	505,514
Less than $50,000	78.3	224,731	8.1	58	42.4	92,326	62.7	317,115
$50,000–$250,000	19.5	55,952	35.2	251	35.6	77,568	26.5	133,761
$250,000–$1,000,000	2.1	5,954	32.8	234	16.2	35,395	8.2	41,583
Over $1,000,000	0.1	326	23.9	171	5.8	12,558	2.6	13,055
Wages per Employee[b]								
Total	100.0	286,963	100.0	714	100.0	217,837	100.0	505,514
Less than $10,000	61.4	176,069	22.3	159	51.1	111,273	56.9	287,501
$10,001–$15,000	16.2	46,539	44.0	314	23.2	50,556	19.3	97,409
$15,001–$20,000	7.6	21,658	21.9	156	12.9	28,178	9.9	49,992
$20,001–$25,000	4.3	12,243	8.1	58	6.7	14,632	5.3	26,933
Over $25,000	10.6	30,454	3.7	27	6.1	13,198	8.6	43,679

Note: [a] Firms included in the table are covered employers with both positive wages paid and positive employment for July 1979 through June 1980. Firm types are defined as follows: Noncharge are firms that had no charges against their UI fund account during FY 1979–80. STC are firms that participated in the STC program sometime during FY 1979–80. Regular UI are firms that incurred charges to their UI fund accounts during FY 1979–80 and who did not participate in the STC program. [b] These data reflect per employee annual compensation per job, rather than total per employee wages across all jobs.

Source: Combined tables from pages 4.3, 4.4, and 4.6 of *California Shared Work Unemployment Insurance Evaluation.* Sacramento: Health and Welfare Agency, 1982.

choices made by different kinds of business enterprises. Using this premise, we may anticipate different kinds of employers among users of STC and regular UI. Table 5.2 shows three comparisons of STC users with other employers in California, from which we may begin to discern profiles of the employers who use STC and those who do not.

Table 5.2 shows generally that STC users are larger, primarily manufacturing concerns that pay above-average wages. Forty-five percent of STC users—versus only 7.9 percent of all California employers—are in manufacturing. Eighteen percent of STC users—versus only 1.7 percent of all California employers—employ over 100 persons. The portrait of STC employers as typically larger is enhanced by a comparison of total employer payrolls. Over 23 percent of STC user payrolls exceed one million dollars compared with only 2.6 percent of all California employers. Finally, STC employers pay higher-than-average wages. Whereas most employers in California paid less than $10,000 per employee in annual wages, most STC employers paid annual wages between $10,000 and $25,000 per employee.

An early concern of STC critics—notably the California Chamber of Commerce and at least one academic economist[5]—was that the program would attract seasonal or weak employers who would drain UI reserves at the expense of healthier employers. Table 5.3 shows the differences between STC employers and two other categories of employers: employers with no charges against their UI accounts (one year without UI benefit charges) and employers with active UI accounts. The purpose for including these comparison groups here is to address the question of the relative financial well-being of STC employers.

The evidence is mixed. At first glance, STC users appear fiscally healthier, at least in relation to the UI system, than other groups of California employers. Among STC employers, for example, we find 24.2 percent with UI reserve balances in excess of $25,000. Even noncharge employers with no recent layoff experience have only one-tenth of 1 percent of their numbers in that category. This finding, however, should be looked at cautiously. As stated earlier, STC users tend to be larger than any other category of California employers, regular UI, noncharge, or total employer population. A $25,000 reserve for a business employing over 100 workers does not guarantee UI solvency. In the event of a substantial layoff or period of STC, such a reserve could easily be depleted.

Fortunately, indicators less dependent on user size are also available. The relative health of STC users is also reflected in the figures on UI Reserve Rate. Even though STC users have about the same proportion of negative-reserve balances as the general population of employers in California (15.3 percent versus 14.4 percent), STC users have a larger proportion of their numbers under a rate of 1 percent than do regular

Table 5.3. Comparison of Employers by UI Charges within Categories of Reserve Balance, Reserve Rate, and UI Tax Rate[a]

Reserve Balance	Noncharge		STC		Regular UI		Total	
	Percent	Number	Percent	Number	Percent	Number	Percent	Number
Total	100.0	286,963	100.0	714	100.0	217,837	100.0	505,514
Lower than −$10,000	0.2	692	6.0	43	5.8	12,603	2.6	13,338
−$10,000 to $0	4.4	12,533	9.2	66	21.7	47,205	11.8	59,804
$1 to $10,000	94.1	269,933	42.6	304	57.2	124,708	78.1	394,945
$10,001 to $25,000	1.2	3,429	17.9	128	9.1	19,743	4.6	23,300
Over $25,000	0.1	376	24.2	173	6.2	13,578	2.8	14,127
Reserve Rate[b]								
Total	100.0	286,963	100.0	714	100.0	217,837	100.0	505,514
Negative Reserve	4.6	13,058	15.3	109	27.5	59,806	14.4	72,973
0–1.0	4.5	12,978	6.7	48	6.0	13,029	5.2	26,055
1.1–5.0	54.8	157,248	67.2	480	45.2	98,411	50.7	256,139
Over 5.0	36.1	103,679	10.8	77	21.4	46,591	29.7	150,347
UI Tax Rate								
Total	100.0	210,320	100.0	707	100.0	212,496	100.0	423,523
0–2.0	31.2	65,602	10.7	76	13.0	27,676	22.0	93,354
2.1–3.0	30.9	64,953	61.7	436	41.1	87,269	36.0	152,658
Over 3.0	37.9	79,765	27.6	195	45.9	97,551	42.0	177,511

Note: [a] Firms included in the table are covered employers with both positive wages paid and positive employment for July 1979 through June 1980. Firm types are defined as follows: Noncharge are firms that had no charges against their UI fund account during FY 1979–80. STC are firms that participated in the STC program sometime during FY 1979–80. Regular UI are firms that incurred charges to their UI fund accounts during FY 1979–80 and who did not participate in the STC program. [b] The reserve rate for a firm is equal to its reserve balance as of June 30, 1980 divided by total wages paid during FY 1979–80.

Source: Adapted from *California Shared Work Unemployment Insurance Evaluation*, Sacramento: Health and Welfare Agency, 1982, p. 4.6.

UI users (see table 5.3). Conversely, fewer STC employers had rates over 5 percent when compared to any other employer type. Finally, a comparison of the UI tax rates of STC users and regular UI users shows that as of California fiscal year 1979, STC users were paying a lower UI tax rate. At least in terms of the California UI system, STC users are fiscally healthier than either regular UI users or the general population of California employers.

One feature of STC users that is particularly ironic is that only 30 public employers out of a total of 4,700 employers have used the program as of the end of 1982. The concern that originally prompted the program—cushioning the impact of the fiscal crisis that Proposition 13 was expected to have on public employees—was negated by the state bailout of local government. Public-sector use is further discouraged by the fact that public employers must directly reimburse the fund for all benefits drawn by their employees because they have not built up reserves through regular contributions based on the payroll as have private employers. Though the STC program might benefit their employees and ultimately the communities themselves, the fiscal crisis they find themselves in leads to a focus on short-range budget balancing, not long-range planning.

As an employer, the state of California has not utilized its own STC program as of the first quarter of 1983. Hiring freezes, attrition, reorganization, and efficiency measures have thus far been sufficient to deal with the fiscal crisis. But state layoffs are looming, and so is the issue of whether to use the STC program to cushion the blow of these layoffs.

Who Shares the Work: Characteristics of Employees

One student of the history of unemployment came to this generalization after reviewing several centuries of policy responses to unemployment:

> They [workers] fear the loss of their jobs but also the competition of the jobless. Given the choice between having fellow workers discharged or sharing limited opportunities with them and thus accepting a reduction in income, relatively few individuals and still fewer labor organizations have willingly accepted the second alternative. Yet there have been times when the labor force has shown a remarkable solidarity.[6]

As we turn from employers to employees, the first question to be considered is the effect, if any, of unionization on STC participation. Because of national labor opposition to shared work (at least until 1981), union reaction to STC in California has been considered particularly significant. Figure 5.2 shows the level of participation by unionized employees.

The only discernible trend across the three and one-half years of STC is the modest percentage of union *employers* participating and the wide

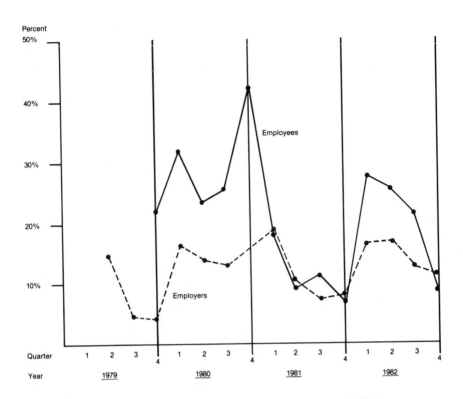

Percent
50%

40%

30%

20%

10%

Employees

Employers

Quarter 1 2 3 4 1 2 3 4 1 2 3 4 1 2 3 4

Year 1979 1980 1981 1982

Fig. 5.2. Quarterly Percent Union Enrollments in the STC Program, 1979–1982 Employers and Employees

Source: The figures on this graph have been tabulated from various issues of EDD, "Work Sharing Activities Monthly Reports."

fluctuation in percentages of union *employees* enrolled. Overall, the contribution made by union worker participation in STC has paralleled that of California as a whole. Approximately 25 percent of the work force of California works under collective bargaining agreements, while an average of 22.2 percent of work sharers were union members. Union *employers,* however, make up 20 percent of all California employers but only 12 percent of STC users. The data again suggest, as with STC use in general, that STC employers are larger firms. Comparatively fewer union employers contribute a similar proportion of workers to the STC population when contrasted to the statewide labor force.

Seniority, frequently seen as a fundamental principle of unionism, has not precluded the use of STC. Obviously some unions viewed sharing work as being advantageous to them and agreed to an employer's plan to enter the program. Shared work and seniority strategies are not directly antagonistic labor-reduction principles, but their coexistence in employment policy does create a tension. Senior workers must, at least temporarily, waive their expectation of full-time work in order for STC to be used. Sixteen and four-tenths percent of STC union representatives mentioned that they feared STC would undermine the seniority principle. However, 47.3 percent of these same officials reported the use of short time without UI prior to utilizing STC provisions, and over 90 percent reported a favorable experience with the state's new employment program.[7]

A willingness to use STC in certain employment situations should not be interpreted as a retreat from long-standing union objectives such as full employment and a secure standard of living but as an adaptation to a major convulsion now occurring in industrialized America. As one business agent remarked: "This is a unique time. There are no other jobs to go to. It's more important to give consideration to being a human being than to contractual technicalities."

What of the nonunion worker? Since it is the employer's prerogative to initiate participation in STC, and only unions have explicit review rights on STC work plans, the profile of work sharers in general, union and nonunion, has implications for such issues as affirmative action.

Table 5.4 summarizes some demographic characteristics of all employees (union and nonunion) who drew benefits under the STC program during 1980, comparing them to regular UI beneficiaries. STC beneficiaries are more likely to be older than 25 years of age than regular UI recipients. The proportion of males and females in the two programs is quite similar. STC benefits go to more members of Hispanic and other ethnic minority backgrounds (with the exception of blacks) than regular UI benefits. Few blacks were STC beneficiaries in 1980.

In general, it appears that STC benefits do reach a greater proportion of ethnic minorities than regular UI benefits (whites make up 48.5 percent of STC employees compared to 59.2 percent of regular UI employees). However, because of the uneven distribution of minority and female employees across California industries, STC provisions do not appear to disproportionately protect those minority group members who otherwise would have lost their jobs.[8] The affirmative action impacts of shared work have been assumed in policy discussions but remain unsupported by California STC experience to date. In practice, protection of the recent gains in employment for minorities and women may be attenuated by layoffs of the newly hired prior to program use. The negligible impact of STC on affirmative action gains could be the result of prior layoffs disproportionately affecting minorities and women. Finally, these affirmative action target groups (particularly females and minority

Table 5.4. Selected Characteristics of Employees Enrolled
in the California STC Program

	STC		Regular UI	
	Percent	Number	Percent	Number
Age	100.0	16,594	100.0	368,735
Total	14.3	2,379	21.4	78,827
Under 25	33.5	5,554	36.2	133,437
25–34	26.5	4,392	18.9	69,601
35–44	25.7	4,269	23.6	86,870
Over 45				
Sex				
Total	100.0	16,594	100.0	368,750
Male	58.4	9,696	61.6	227,225
Female	41.6	6,898	38.4	141,525
Race				
Total	100.0	16,594	100.0	368,725
White	48.5	8,055	59.2	218,250
Black	4.5	749	9.9	36,600
Hispanic	34.3	5,699	27.6	101,850
Other	12.6	2,091	3.3	12,025
Occupation[a]				
Total	100.0	3,076	100.0	368,761
White Collar	18.0	555	30.3	111,750
Blue Collar	69.3	2,132	40.6	149,807
Service	2.9	89	8.3	30,775
Farm and Labor	9.8	300	20.7	76,429
Industry				
Total	100.0	14,402	100.0	368,735
Construction	2.3	334	14.9	55,113
Manufacturing	80.3	11,565	31.7	116,763
Trade and Services	11.4	1,637	33.1	121,892
Other	6.0	866	20.3	74,967

Note:[a] Occupational data for STC claimants are based on an unexpanded
20 percent sample, while the regular UI figures are from an expanded
sample of the same proportion.

Source: Adapted from Shared Work Unemployment Insurance Evaluation
(Sacramento: Health and Welfare Agency, 1982).

group members) are most concentrated in the public sector. When the
time finally does come to cut public labor costs, STC may yet prove to
be protective of minority and female employment.

Financial Consequences of the STC Program

The California Employment Development Department, using elaborate
scenarios constructed partly on empirical data and partly on economic

assumptions, has analyzed the financial impact of STC on user employers, employees, and the state. That analysis, based on the best available estimates, showed that employers (exclusively private firms in the sample) saved $16 per employee per week by using the STC alternative rather than layoffs and regular UI. The bulk of these savings came from reducing the turnover costs associated with rehiring and retraining employees.

The average number of employees placed on shared work was 25.5, and the average duration of shared work was 13.3 weeks. The average time reduction was slightly more than one workday in five (20.8 percent).

Employees benefited differentially, depending on whether or not they were likely to have been laid off had STC not been used. Senior employees, relatively immune to layoff, were found to lose 10 percent of full-time weekly income on the average. Employees likely to have been laid off received more than twice (205 percent) the amount of their regular UI benefit. Overall, the average worker maintained 92 percent of full-time wages and benefits while working, on the average, a four-day week. Thus, both employees and employers are seen to have gained under the program, with the possible qualification that the income of senior, layoff-immune workers is redistributed to junior workers.

The EDD analysis showed that STC does cost the state slightly more than regular UI primarily because of (1) reduced income tax revenues (STC reduces all workers' income and therefore their tax rates), and (2) increased UI benefits and administrative costs. One caveat is that administrative costs were accrued during the "breaking-in" period of STC, for which California had no other American models to emulate. Administrative changes and streamlining have continued in California.[9]

The decrease in income tax revenues—primarily federal taxes—amounts to an average of 10.7 percent of per-worker weekly tax receipts. Workers drawing STC benefits cost the state 15.8 percent more than their laid-off counterparts, according to EDD figures.[10] EDD projections, however, suggest that the surcharge levied against negative-account STC users, coupled with limited program use, will protect the solvency of the state's UI fund.

A word is in order about the simulated costs described above. Due to time and budgetary constraints, the California evaluation team was unable to collect a usable sample of firms actually using layoffs with which to compare STC employers. Instead, simulations were run on the approximately 300 "STC firms" for which extensive economic data had been compiled. The financial figures summarized above reflect *simulated* work sharing across these firms and *simulated* layoffs. Actual data on wages, fringe benefits, tax schedules, and other parameters were incorporated into the scenarios and the models manipulated to estimate per-week costs and full-term impacts.

First, certain assumptions were necessary to construct such simulations. They were:

1. *Depth of labor reductions.* The layoff reductions used for the simulation were equivalent to the work-time reductions reported for STC. If a firm reported an average work-time reduction of 20 percent with shared work, then 20 percent of the firm's work force was assigned the status "laid off" for the layoff simulation.

2. *Duration of labor reductions.* Under both scenarios, STC and layoff, the duration of reduction was the same as reported for STC. The full period of labor reduction for both cases was equivalent.

3. *Layoff criteria.* Assignment to a "laid-off" status under the layoff simulation was based on two sets of ranking: seniority and wage rates. Within a given firm, workers were first ranked by length of tenure with their STC employer and then further ordered by wage rate. Wage rate was used as a partial measure of skill, function, and productivity. The least vulnerable workers in this scenario, therefore, were the longest tenured *and* highest paid. Those laid off in a simulated layoff were the most junior *and* lowest-paid workers.

Any or all of these assumptions might be called into question should compelling empirical data surface on depth and duration of layoffs vis à vis STC. Additionally, layoff criteria are always multifaceted and influenced by specific employment contexts. For the sake of an initial estimation of the financial impact of STC, however, none of these assumptions is unreasonable. However, they should be tested when more complete data become available.

Second, as the California report stated, the analysis presented was based upon work groups within firms. Frequently, only a designated work-sharing group or groups, rather than the entire work force of an employer, participated in STC. Since UI tax rates are assessed on an employer's gross payroll, any depletion of an employer's UI account stemming from STC will be compensated by a "leveraging effect" due to the higher tax rate being applied to all workers, not just those sharing work. The effect should insure UI reserve solvency.

Third, California's financial impact scenarios were not full-blown cost-benefit analyses. Certain effects, such as differences in the patterns of consumption of fully retained workers, laid-off workers, and work sharers, and the resulting changes in state revenue (sales and excise taxes, for example) were omitted. Any attempt to include these on the basis of current data would have required highly speculative estimates. These effects, nonetheless, are thought to be important derivatives of the STC program.

The social costs of unemployment, including mental health effects, crime, bankruptcy, and others, were not included as costs to the state. Many social science studies have already established the very real presence of these miseries among the unemployed, but their specific cost estimation continues to be legerdemain at best.

STC results in job-service savings when compared with regular UI. Work sharers, for example, are not required to utilize employment service and job placement procedures. The savings involved in a programmatic exemption of this sort were also not included.

Finally, the financial repercussions on the UI system of neglected STC claims were not included. Since the loss of income is substantially less under STC for those who would have been laid off, failure to file for benefits could be substantial. A survey of 265 STC employees revealed that 22 percent reported neglecting at least one weekly UI entitlement during their experience with the program, despite a mail claim system in California that eliminates the need for weekly visits to UI field offices. Seven and six-tenths percent missed three or fewer claims, and 14.4 percent failed to collect four or more claims. Though these figures hint at savings for the California UI reserve, such negligence on the part of claimants is probably variable and subject to many eccentricities in the economy, individual work situations, and UI wage replacement levels, and not something to bank policy on.

The specific financial analysis undertaken by the California evaluation of STC was purposefully delimited and reflected the gains and costs associated with direct factors for which reasonable financial estimates were available.

Recent Legislative Changes

Since July 11, 1978, STC has been an employment policy experiment in California. The program was sunset legislation, renewed in 1979 and extended in 1983 by Senate Bill 57 until 1986.

Senator Bill Greene introduced the bill to accomplish two important changes—in eligibility and workweek definition. Eligibility was originally limited to 20 weeks in a 52-week period. The new law permits 26 weeks of STC, but these weeks can no longer be spread out over the 52-week period. However, this 26-week limit is effectively extended to 52 weeks (although separate applications for each 26-week period are required) whenever the civilian work force unemployment rate exceeds 7.5 percent. California's unemployment rate exceeded 7.5 percent from September 1974 through November 1979 and passed the 7.5 percent mark again in September 1981. Thus, the state has surpassed the 7.5 percent trigger for 58 out of the last 102 months, as of March 1983. With full employment now considered to be about 5 or 6 percent, continuous use of STC promises to be a part of the next decade.

A second significant provision of the new law changes the definition of a full workweek from "an individual's normal hours or number of days a week" to "an individual's normal weekly hours of work or 40 hours, whichever is less." Normal working hours, therefore, can never be more than 40 hours a week, but they can be less. The STC program acknowledges the economy's move toward shorter workweeks.

Finally, the new version of the law provides that if an employer has a negative reserve account balance on June 30 of two consecutive years, the employer would pay contributions equal to the amount of benefits paid under the program during the prior year, and would provide for a refund or credit to the employer for amounts collected in excess of this amount during the prior year.

CONCLUSION

California's STC program is quite likely the first major innovation in American unemployment policy since the New Deal, certainly since the work-force programs of the 1960s. As such, it has generated interest among legislators, analysts, employers, and unions around the country. The program also draws attention to some broader social issues regarding employment and work.

First, STC represents a new phase of the state's involvement in the private employment relationship. Second, it introduces a shorter, more flexible workweek, which clouds the once clear distinction between being fully employed and being out of work. Third, it directs attention to the complexity of using employment policy to reach groups that traditionally have been discriminated against in the labor force. Fourth, it suggests a redefinition of leisure. Finally, STC comes at a time when the whole country, as well as California, is undergoing a major industrial and economic transformation. Traditional industrial sociology, with its emphasis upon management and labor inside the factory system, cannot adequately address these new and changing problem areas.

The STC program demonstrates one way in which the state and the economy interact. At a time when state budgets are pressed and unemployment is high, STC may be more economical than direct job creation for employment stabilization. The California program's development, however, was not straightforward. It was inaugurated to benefit potential victims of public-sector layoffs but has evolved into a more permanent fixture of state employment policy available to the private sector. While this development may seem fortuitous, it is important to watch where the program goes from here. Its increased use over the last four years, especially by large employers, suggests a new kind of partnership between the state and private business. An important issue for students of the new industrial sociology is the role of the state in private-sector employment and how that role may vary with large and small firms.

The trend toward shorter work hours is not new. In 1887, Samuel Gompers said, "As long as we have one person seeking work who cannot find it, the hours of work are too long." The extent to which the STC program is a part of the general trend toward shorter hours may be disconcerting because it presents no new employment opportunities. The issue remains as in Gompers' time the distribution of limited employment opportunities among an expanding work force. STC does not expand employment opportunities but is rather a response to contracting opportunities. Policymakers must look elsewhere for programs that create new employment opportunities.

Recent theories of labor markets are also relevant to the study of the STC program, especially its impact on affirmative action. Conventional labor-market theory has treated labor as an undifferentiated mass, homogeneous and mobile. Theories of segmented labor markets posit that certain groups, minorities and women in particular, have historically occupied less valued positions in the labor market. Their jobs are the low-skilled, low-paid, part-time employment on the fringes of the labor force. They are also those most easily replaced when they leave an employer.

Another labor-market segment consists of the valued employees with the "good" jobs—the higher-skilled, higher-paid, full-time workers. These workers are more difficult for employers to replace. Historically, this segment is dominated by white males. Since employer initiative determines which employees will participate in shared work, the jobs considered valuable are defined by the employer. This leaves open a theoretical question: does employer-initiated STC aid certain labor-market segments and not affect other segments for which it was originally intended? Based on California's experience, this question remains unresolved.

Is it an increase in "leisure time" when workers have an extra day off each week without pay but with STC benefits? Past perspectives on leisure defined it as a luxury commodity of a prosperous postindustrial society. The question then was, how shall we constructively consume our leisure time? But when the extra day off each week results from an austerity move on the part of an employer, shall *that* be considered increased leisure?

Policymakers in other states will undoubtedly look to the California experience for guidance. While the STC program raises broad and important questions for policy analysts and industrial sociologists (such as those we have addressed here), its actual impacts are still relatively modest. One thing is certain: work in America is changing. STC is likely to be an important part of that change for the foreseeable future.

NOTES

1. These figures and rules are taken from the *1982 Employers' Guide* (Sacramento, Calif.: California Employment Development Department, 1982).

2. California Employment Development Department, *Report to the Governor on Labor Market Conditions* (Sacramento, Calif.: California Employment Development Department, 1983).

3. The pattern of participation of large employers may reflect EDD efforts to promote the program (see below) as well as a leveling of unemployment in California during 1981. The proportion of large employers (over 100 workers) participating shows an increase within each year, with the greatest gains in 1982:

STC Employers Enrolling over 100 Workers Each Quarter
1980–1982

	1st Quarter		2nd Quarter		3rd Quarter		4th Quarter	
	%	(n)	%	(n)	%	(n)	%	(n)
1980	4.2	(5)	8.2	(35)	10.4	(22)	9.9	(13)
1981	5.4	(9)	6.3	(10)	6.0	(7)	7.9	(38)
1982	10.6	(63)	12.0	(90)	14.8	(95)	12.2	(83)

Source: Compiled from various reports on work-sharing activities by the California Employment Development Department.

4. State of California Employment Development Department, *Report to the Governor on Labor Market Conditions and Work Sharing Activities*, and other reports. California reports STC enrollees as initial UI claimants, but also as employed persons. Like regular UI clients, not all STC enrollees collect their benefits (see below). Therefore the 15 percent figure cited is inflated to its upper limit.

5. Daniel Hammermesh, "Unemployment Insurance, Short-Time Compensation and the Workweek," in *Work Time and Employment: A Conference Report*. Special Report No. 28 (Washington, D.C.: National Commission for Manpower Policy, 1978).

6. John A. Garraty, *Unemployment in History: Economic Thought and Public Policy* (New York: Harper & Row, 1978).

7. State of California Health and Welfare Agency, *California Shared Work Unemployment Insurance Evaluation* (Sacramento, Calif.: Health and Welfare Agency, 1982).

8. Ibid.

9. Because the STC program is a different kind of employment security program than regular UI, a number of administrative changes have occurred over the five years the program has existed in California. Some of the more substantial changes are: (1) in cases where large employers elect to use the program, EDD field office personnel will register employees for the program at the work site rather than have each individual worker come to a field office, thereby saving time and expense; (2) the payment of STC benefits was centralized in July 1982 (plan approval has been centralized since the program's inception); (3) work-search requirements were dropped early in the program's development for not being in the spirit of shared-work legislation; and (4) a reduced workweek can include a week where no actual services were performed, but a week that included some vacation days and some "shared work" days when the employer did not offer work.

10. State of California Health and Welfare Agency, *California Shared Work Unemployment Insurance Evaluation*.

6.
Arizona, Motorola, and STC

Robert St. Louis

The state of Arizona was hard hit by the 1973–75 recession. Whereas the national unemployment rate peaked at 9.1 percent in June of 1975, Arizona's rate peaked at 13.6 percent in that same month. The fact that Arizona was especially hard hit is not surprising. The major industries in Arizona are tourism and construction, both particularly vulnerable to downturns in economic activity. Problems in the copper and electronics industries further contributed to Arizona's unemployment problems. The length and severity of this recession led many employers and government agencies to explore procedures to prevent a recurrence of the layoffs that accompanied the recession.

The first dialogue between a private employer and the state of Arizona occurred in October 1977, when staff members from Motorola contacted the Arizona Department of Economic Security (DES). Motorola's management had developed a broad plan for stabilizing the company's employment, and work sharing was a major part of that plan.

For numerous reasons, Motorola's management believed that work sharing would be acceptable to employees only if unemployment insurance (UI) benefits were paid to those employees whose hours were reduced as a result of work sharing. Motorola pointed out that even though the employer pays the UI tax, many employees feel unemployment benefits are an earned fringe benefit. Hence, employees would feel cheated if the company avoided having to pay UI benefits by reducing the number of hours each employee works rather than the number of employees. The company also felt that partial unemployment insurance benefits were needed to make work sharing acceptable to the senior employees whose hours would not normally be reduced by a layoff.

The department was very skeptical of Motorola's initial overture. In fact, the first question raised by the department was why Motorola needed cooperation from DES. It clearly was within Motorola's power to reduce the hours of its employees, and it clearly was within its power to partially compensate employees for those lost hours. So why did the department have to become involved?

Simulating the 1973–75 Recession

To determine whether the department needed to be involved, a simulation experiment was conducted. The experiment consisted of translating Motorola's actual layoffs during 1973 through 1977 into equivalent work-sharing reductions and observing the amount of unemployment insurance benefits that Motorola would have had to pay those employees whose hours were reduced. An assumption of the experiment was that the total reduction in hours would have been the same under work sharing as it had been under the layoffs that actually occurred. The results of the simulation experiment were very informative. They showed that:

1. Had Motorola used corporate funds to partially compensate those employees whose hours were reduced as a result of work sharing, a severe (3.5 million dollar) cash-flow problem would have arisen at a time when the corporation could least afford it.

2. If Motorola were to set up a private account to partially compensate workers whose hours were reduced as a result of work sharing, that account could not be funded from reductions in the firm's statutorily required unemployment insurance account.

The first result was not surprising. A firm would reduce hours only during bad economic times, and during bad economic times a firm doesn't need to incur added expenses. A consequence of the first result, however, was that a firm almost certainly would not be able to partially compensate workers whose hours were reduced as a result of work sharing unless the firm was able to effect a reduction in some other cost to offset the partial compensation or short-time benefits.

Initially the department assumed that work sharing would reduce Motorola's unemployment insurance taxes by an amount large enough to allow Motorola to fund its own account for making short-time benefit payments to workers with hours reduced as a result of work sharing. The department felt this would be true because unemployment insurance taxes are "experience rated." Firms with frequent layoffs (bad unemployment experience) pay taxes at higher rates than firms with infrequent layoffs (good unemployment experience). Because work sharing would greatly reduce layoffs, the department felt it would reduce unemployment insurance taxes by an amount approximately equal to the short-time benefits that Motorola would pay its workers.

However, the simulation experiment showed the department had overlooked three factors: the size of the reserve-account balance, the time lag required to recoup short-time benefits, and the pooled-risk principle.

In Arizona, an employer's tax rate is determined by the employer's reserve ratio (the employer's reserve-account balance divided by the employer's average taxable payroll). The higher the reserve ratio, the lower the tax the employer must pay. When the reserve ratio is 13 percent or more, the employer qualifies for the minimim tax rate. What the department initially overlooked was the fact that Motorola's statutorily required unemployment insurance reserve account balance would not be reduced as a result of work sharing. Thus, if the department could not or would not pay short-time unemployment insurance benefits, Arizona employers wishing to do so would have to maintain two reserve accounts: one with the state for the payment of statutorily required unemployment insurance payments and one within the firm for the payment of short-time compensation (STC) benefits.

The department also underestimated how long it would take a firm to recoup its STC payments through a reduction in its unemployment insurance taxes. Because of the manner in which tax rates are set, a reduction in the tax rate lags an improvement in unemployment experience by about 18 months. The department was aware of this lag, but overlooked the fact that the reserve-account balance continues to increase until the minimum tax rate is achieved. This means that only part of the savings from improved experience are returned to the firm initially. In Motorola's case, $900,000 would have gone to increasing the statutorily required reserve-account balance (because without work sharing Motorola's long-run reserve ratio would have been less than 13 percent), and it would have taken nearly 10 years to recoup the remaining 2.6 million dollars that would have been paid out during the recession.

The simulation experiment also showed the department that it had initially overlooked the pooled-risk principle. That is, to provide a given amount of protection against running out of funds, an employer would have to maintain a much larger reserve-account balance when acting alone than when acting in concert with other employers. Thus the simulation experiment convinced the department that if an employer were to set up a private account to partially compensate workers whose hours are reduced as a result of work sharing:

1. The employer's statutorily required reserve-account balance would increase, not decrease (unless the employer's reserve ratio already was 13 percent).
2. The employer's unemployment insurance tax would decrease (unless the employer already was paying the minimum rate) but by less

than the full amount of the short-term benefits paid (because the statutorily required reserve-account balance would increase).

3. There would be a lag of at least 18 months before the short-term benefits would begin to be recouped through lower unemployment insurance taxes.

4. To provide protection against running out of funds during bad economic times, the employer would have to maintain a separate private reserve account at least as large as the statutorily required reserve account.

Perhaps the most important result of the simulation was that since the total weeks of unemployment were not changed, it led DES officials to the view that a short-time unemployment insurance program was simply offering employers an option that they heretofore did not have with respect to how their unemployment insurance contributions would be paid out. The simulation was completed in March of 1978.

Studying Short-Time Unemployment Compensation (STC): A Department Goal

Given the results of the simulation, Henry Haas, the UI program administrator, began to search for funds to further study the issue. During 1978, proposals to study STC were submitted to both the U.S. Department of Labor and the National Commission on Unemployment Compensation. Neither of those bodies was able to fund the proposals, so Haas began to look inside Arizona for funds.

Largely because of Haas's interest and initiative, the DES Advisory Council approved the study of STC as a department goal in May of 1979.

The Advisory Council is made up of three business, three labor, and three public representatives. Eight members of the council felt STC deserved further study, and one labor representative did not. The labor representative's opposition was based on his belief that the department's limited resources could be better spent on other projects.

In addition to approving the goal, the Advisory Council requested that progress reports be submitted at all subsequent council meetings. The council also urged the department to work closely with business representatives, labor representatives, and legislators when designing the study. Because paying unemployment insurance benefits to persons who are working appeared to be radically different from the historical goals of the unemployment insurance program, the council warned that initial reactions to STC could be very negative unless the concept were carefully explained.

The Arizona Survey

Before developing a specific proposal, meetings were held with legislators, business leaders, and labor leaders to explain STC and to ascertain the types of questions they had concerning STC. As the Advisory Council had anticipated, initial reactions were generally negative. After discussion and clarification, however, almost every person contacted agreed that the concept deserved further study.

Arizona Senator James Mack had independently submitted a bill to appropriate funds to study STC, and meetings were also held with Senator Mack and members of the appropriations committees in both the House and Senate. During those meetings it was emphasized that

1. STC offers potential advantages to both employers and workers.
2. STC offers employers and employees a new option with respect to how unemployment insurance contributions are to be paid out in benefits but should not significantly increase the total amount of benefits paid.

On March 19, 1980, Senator Mack's bill was approved at a joint meeting of the Senate and House appropriations committees. The vote was 24 for and four against, with two absent. The four opponents of the bill were neither all Republican nor Democrat, neither all prolabor nor pro-business. They simply didn't feel the concept merited a $20,000 study. They weren't necessarily negative toward the concept; they just weren't convinced it had any potential advantages.

Following final passage of the bill, meetings were held with local businessmen and labor leaders to determine what they felt were the issues surrounding STC. The major issues that surfaced were:

- Are STC benefits needed to make work sharing acceptable to employers and employees?
- Could the program be abused by employers or employees?
- How long should employees be able to draw STC benefits?
- What should be the maximum reduction in hours permitted in an STC program?
- Should employees drawing STC be required to seek full-time work?
- How many persons favor trying an STC program?

Following a discussion of STC and the proposed study, the Arizona Association of Industries (AAI) contributed $10,000 to support the study. Organized labor did not financially support the study but agreed to participate in it.

A decision was made to design the study to determine the attitudes of employers, union leaders, and workers toward a pilot STC law. Because only limited funds were available, it was felt that the most the study could accomplish would be to determine: (1) whether anyone wanted an STC law; and (2) the most desirable form for such a law. Results from the survey were published in 1981 and are available upon request to the Unemployment Insurance Administration of the Arizona DES. Highlights of the report were:

- 67 percent of union representatives favor temporarily changing the UI law to allow STC to be tried in Arizona.
- 52 percent of employer representatives favor temporarily changing the UI law to allow STC to be tried in Arizona.
- 70 percent of employees work for employers whose representatives favor temporarily changing the UI law to allow STC to be tried in Arizona.
- 81 percent of employees favor temporarily changing the UI law to allow STC to be tried in Arizona.
- A shared-work program must include STC if it is to be acceptable to union representatives, employer representatives, or workers.
- Shared-work programs that last 26 weeks or less are preferred to those of longer duration.
- Employers who wish to participate in an STC program should be required to certify that the reduction in work hours will be in place of a layoff.
- Employers participating in an STC program should not be allowed to increase the number of employees in the affected unit(s).
- Employees participating in STC programs should not have their hours reduced by more than 40 percent.
- Employees participating in STC programs should not be required to seek full-time work to qualify for benefits.
- Moonlighting income should not be deducted from STC benefits.
- The UI taxes of employers who, for whatever reasons, do not participate in the STC program should not increase because of STC.
- Negative-balance employers who participate in an STC program should pay a surtax.

Three of those concerns were almost universally endorsed by employees, employers, and employee representatives:

1. STC programs must be of relatively short duration if they are to be acceptable to the majority of employees.
2. STC programs should not reduce employee hours by more than 40 percent.
3. Non-STC employers should not be forced, through their UI contributions, to subsidize STC employers.

Also of interest are the rankings of the potential advantages of STC.

Potential Advantages

Rank of Potential Advantages	Union Representatives and Workers	Employer Representatives
1	More equitably distributes the burden of unemployment during economic downturns. (Many employees have their hours reduced as opposed to a few employees losing their jobs completely.)	Maintains a high level of morale, productivity, and quality by keeping the work force intact.
2	Maintains the fringe benefits of those who would otherwise be laid off.	Retains a skilled work force, enabling the employer to respond more quickly when business improves.
3	Preserves job skills of workers who would otherwise be laid off.	Reduces hiring and training costs when economic conditions and production pick up.

While both employers and employees saw more potential advantages than disadvantages in STC, employees were concerned that under STC they would have smaller accruals to pension and other wage-related fringe-benefit programs. Employers feared that it might be more difficult to schedule work under an STC program.

Formulation of a Law

With the study as background, it was possible to begin consideration of the desirability of formulating an STC law. A number of issues arose during this process. The issues can be grouped into two general categories: concept issues and implementation issues. There were five basic concept issues:

1. What will be the impact on the UI trust fund?
2. What will be the impact on unemployment statistics?
3. What will be the impact on employers' tax rates?
4. Will the program result in higher administrative costs?
5. Since employers can work share on their own, why should the state become involved?

Three assumptions were made in examining these issues. Arizona's STC program would parallel the California experience of limited early participation and slow growth. Arizona's economy would slowly but steadily improve in 1982 and 1983. Arizona's legislation would be limited to a two-year pilot program. In California during 1980, less than one-half of 1 percent of all weeks compensated had been STC claims. In Arizona at the beginning of 1981, the unemployment rate was about 6.5 percent, the insured unemployment ratio was about 2.75 percent, and the UI trust fund balance was about $260 million, which seemed to be adequate.

Given the above information, it was easy to demonstrate that the impact on the UI trust fund and unemployment statistics would be very slight during the two-year pilot program. It was realized, however, that both of those issues would have to be addressed again as the pilot program neared completion. By that time it was hoped that the Department of Labor would have determined how to include STC weeks in its statistics and that either the California evaluation or results from Representative Patricia Schroeder's proposed bill in the Congress would have shown the impact of STC on the UI trust fund.

By imposing a surtax on negative-balance employers who chose to participate in the STC program, it was possible to demonstrate that there would be no long-run impact and very little, if any, short-run impact on employers who chose not to participate in the STC program. At the same time, efforts—not entirely successful under all circumstances—were made to formulate the law in such a way that negative-balance employers were not penalized through the surtax for more than the additional cost to the fund caused by their use of STC.

By computerizing the majority of the STC program operations and eliminating the necessity for claimants to visit local UI offices, it was possible to demonstrate that five STC claims could be processed for approximately the same cost as one regular claim. Since STC claimants are not required to register with or otherwise use the job service, it was thought that overall administrative costs might actually be lower for the STC program than for the regular program.

And after observing the extent to which work sharing has been utilized in the United States and European countries, it was decided that the payment of STC benefits probably encourages the use of work sharing. Primarily because of the potential advantages to society from work sharing, the department proceeded to draft an STC law.

When drafting the law, six major implementation issues arose. They were:

1. Must STC claimants meet the able, available, and active work-search requirements?

2. How should the amount of the STC benefit be determined?
3. What constitutes an STC waiting week?
4. What are the requirements for an STC plan?
5. How should employers be charged for STC benefits?
6. Should a distinction be made between permanent and temporary reductions in hours?

To require claimants to actively seek full-time work during their off hours (and accept it if offered) would defeat a major purpose of the STC program (to allow employers to retain their skilled work force). Therefore, it was decided that STC claimants need not meet the able, available, or active work-search requirements.

The next major issue concerned the amount of the STC benefit. Should it be determined on the basis of hours worked or dollars paid? How should sick days, vacation days, overtime hours, moonlighting income, and retirement income be handled? The decisions reached were: (a) to ignore moonlighting income; (b) to define the claimant's normal weekly hours of work to be those normally worked or 40, whichever is less; (c) to define the STC benefit payable to be a pro-rata share (based on hours compensated and not hours worked) of a claimant's maximum weekly benefit amount; and (d) to deduct retirement income from the STC benefit.

The third issue concerned the waiting week. Should STC claimants be ineligible for benefits until they have lost the equivalent of their normal weekly hours of work or their normal weekly pay? Should an STC waiting week substitute for a regular waiting week and vice versa? The decisions reached were: (a) there should be a waiting week; (b) the waiting week requirement would be met by any week in which the claimant experienced a reduction in hours in accordance with an approved STC plan; and (c) a regular program waiting week would substitute for an STC program waiting week and vice versa (i.e., only one waiting week need be served).

Another major issue concerned the requirements for an STC plan. Questions surrounding this issue were: Should there be maximum and minimum allowable reductions in hours? Must the hours of all workers listed on a plan be reduced each week the plan is in effect? Can employers add workers to a plan after it is approved? Should the plans have an expiration date? For how long should workers be allowed to draw STC benefits? And should the department be allowed to deny plans of employers whose accounts are delinquent? The decisions reached were: (a) to be eligible for STC benefits, a worker's compensated hours must be reduced by at least 10 percent but no more than 40 percent; (b) employers need *not* specify in advance the reduction in employees' hours; instead the employer should report for each week the number of hours for which each claimant in the plan was compensated (compensation includes wages, sick pay, and vacation pay); (c) the hours of workers listed in an STC plan may or may not be reduced—the plan merely identifies workers who both the employer and the collective bargaining agent (if any) agree may have their hours reduced; (d) workers cannot be added to a plan after it is approved—instead a new plan may be submitted for the employee(s) to be added (this allows collective bargaining agents to be aware of which employees are listed in shared-work plans); (e) each plan must have an expiration date that is no later than 12 months from the date the plan is submitted; (f) workers should be allowed to draw STC benefits for no more than 26 weeks in a benefit year; and (g) DES should not deny a plan unless the shared work is not in lieu of layoffs, one or more employees listed on the plan do not have sufficient prior work experience with the employer, or the plan is not acceptable to the relevant collective bargaining agents (if any).

The fifth issue concerned how employers should be charged for STC benefits. More specifically: (a) should base period employers be charged? and (b) should noncharging be permitted? The decision made was that STC benefits should be charged against employers' accounts in exactly the same manner as regular benefits.

The last major issue concerned whether a distinction should be made between hour reductions that employers feel are temporary and those they feel are permanent. The decision was made that no distinctions should be drawn. The feeling was that employers frequently don't know whether a cutback will be temporary or permanent. Moreover, since workers can't draw STC benefits for more than 26 weeks, those with low seniority have a strong incentive to seek full-time work elsewhere during their off hours if they view the cutback as permanent. It was also decided that claimants would be disqualified from receiving regular benefits if they quit the STC employer to look full time for work.

Implementation of the New Law

A law was drafted and passed in April 1981 but did not become effective until January 1982. Implementation was deliberately delayed to allow the UI administration time to develop an administratively efficient procedure for paying benefits.

Because of the opportunity for careful planning, implementation of STC in 1982 was effected smoothly and without significant problems. Utilization of the new legislation exceeded expectations, and 527 plans covering 25,152 workers were approved during the first 10 months of the year. During the same period only seven plans—all from small employers—were disapproved, all for technical reasons. One plan, for instance, failed to comply with the criteria for approval because it covered a single employee.

During the first 10 months of operation, manufacturing firms accounted for approximately half of the participating firms but some 85 percent of the STC employees. Preliminary data indicate that work reductions have averaged about 20 percent, with the average STC benefit amounting to $24.30 per week. STC benefits for the first 10 months of 1982, the period for which data is available, have totaled $1,573,581.

Women accounted for 57 percent of all STC claimants as opposed to 30 percent of regular UI users. The comparable figures for nonwhites were 31 percent for STC and 27 percent for regular UI. Since a single employer, Motorola, accounted for about 40 percent of STC activity, and since Motorola has an above-average percentage of female employees, it is not possible to assess the degree to which these percentages represent evidence that the Arizona program had an impact on affirmative action.

DES has not yet attempted any comprehensive evaluation of the STC program, partly because it has waited to conduct such an evaluation in cooperation with the U.S. Department of Labor, which is mandated by Congress to make evaluations of state programs.

Despite any formal evaluation, however, there is other evidence that the program has succeeded in Arizona even beyond the expectations of its proponents. Perhaps the most dramatic evidence of this fact came in March of 1983 when emergency legislation making the program permanent (the original law had created a two-year trial program) was passed without a single dissenting vote in either the House or Senate.

The new legislation also provided that the 26-week-per-year limit for the duration of an STC plan would be automatically removed during periods when the insured unemployment rate reaches 4 percent (the "insured rate" is normally approximately half of the regular unemployment rate). This means that in periods of severe recession, Arizona employers will be able to use STC continuously. The only statutory (but inapplicable) limitation on the amount of STC benefits that an individual worker can receive is the provision that the total amount of STC benefits paid to a claimant in a given year may not exceed 26 times the regular UI weekly benefit. This maximum, of course, could never be reached (unless an employee had a minimal work history), since work reductions under STC plans are limited to 40 percent of the normal workweek (two days), and even with reductions of two days for 52 weeks (104 days) a worker would draw less than the regular maximum of 26 full weeks (130 days). The legislation also recognizes the growing prevalence of a less-than-40-hour standard workweek—it defines a week as 40 hours or normal hours, whichever is less.

The unanimous vote in the Arizona legislature making STC a permanent feature of the state's employment security system reflected what appears to be almost universal satisfaction by the users—including employers, employees and unions—of the program.

A *Wall Street Journal* article on April 7, 1983, described the reactions of Motorola's 9,000 employees to STC work reductions. One woman employee was quoted as saying: "It was a great idea. Our girls are highly trained, and it's hard to replace them. When we lose them, we don't get them back and that is a serious loss."

The *Wall Street Journal* reported that Motorola management found that productivity during the period of reduced hours remained high, in sharp contrast to the experience of the company during previous periods when layoffs were utilized. During the recession of 1975, the employee quoted above had been "bumped" from her job by a more senior worker, and she in turn had bumped another worker less senior than herself. She is quoted as recalling: "It was disastrous. I was put in a department I knew absolutely nothing about. By the time I knew what I was doing, I was laid off."

In the same report, C. F. Koziol, Motorola's corporate director of personnel administration, is quoted as saying: "The results have been absolutely gratifying to us. We estimate that we were able to salvage over 1,000 jobs. The ability to reinstate full production schedules gives us a tremendous advantage over employers that had to go to layoffs."

Motorola was perhaps inclined to look favorably on the program since the company was a prime mover in the development of STC in Arizona. Other employers, however, were almost equally enthusiastic. William C. Huston, director of human resources at Dolphin, Inc. in Phoenix, reported:

> Here at Dolphin we have found the [STC] program of great benefit in allowing us to retain a trained work force in the face of the economic situation. Should things pick up in the near future, we would be able to shift into high gear overnight, giving us a jump on our out-of-state competitors most of whom have laid off a large percentage of their work force. . . .

Robert M. Blocher, director of corporate planning and services of Mountain Mineral Enterprises, Inc., in Tucson, echoes the finding:

> [We] began participation in the Arizona . . . [STC] program in May of 1982. We currently have 143 employees who have been affected by our having to reduce the work week from 40 hours to 28 hours. This 30 percent reduction . . . has been necessary for us to hold our work force during the prolonged period of reduced business. The shared work program has not only been very helpful to the company and its employees but we feel it has benefited

the community by helping to keep the rate of unemployment from going even higher.

While only a small percentage of Arizona's work force is unionized, preliminary reaction to STC seemed to indicate that unions were as pleased with the program as employers. William E. Coons, president of Local 184 of the International Chemical Workers Union (ICWU), said: "As President of the ICWU Local 184 and on behalf of the union members, I would like to extend support to the [STC] program. We believe this is a good program and it is very effective."

Final evaluation of the first year of STC in Arizona seems likely to reveal an anticipated bonus—but one of unexpected and surprising magnitude. Approximately 80 percent of regular UI claimants were required to use DES's Job Services at a cost that averaged $212.75 per placement in 1982. Placement is the raison d'être of the Job Service and it is a service not specifically required by STC recipients. If not for work sharing, it is assumed that the 25,152 employees participating in STC during the first 10 months of the program would have resulted in approximately 5,030 full-week layoffs, resulting in 4,024 additional claimants processed by the Job Service. This potential $800,000 reduction in Job Service costs is given additional significance by the fact that during this ten-month period total benefit payout amounted to only slightly more than $1.5 million and total administrative costs to less than $0.3 million. Clearly such potential savings cannot be ignored when attempting to evaluate the desirability of an STC program—especially since it is reasonable to assume that after this initial start-up period the program can be administered more efficiently.

7.
Oregon Tries the "Workshare" Idea

Donna Hunter

In the spring of 1980 the economy of Oregon was hit particularly hard by the credit crunch recession. Both economically and psychologically, the recession caused more havoc in Oregon than in the average American state. After booming growth in the 1970s, Oregonians had a singularly positive sense of their state as strong and self-reliant, a continuing American frontier where the good life was attainable without the environmentally destructive industrial growth that had characterized so much of the Northeast and Midwest.

The 20 percent interest rates that accompanied, or perhaps caused, the recession had a direct effect on Oregon. Housing, followed closely by automobiles, is the sector of the economy most sensitive to interest rates, a response that is readily understood when one considers that an increase of just a few percentage points in the mortgage rate can literally double the ultimate cost of a home. The precipitous national decline in housing had an immediate effect on Oregon's economy, which is heavily dependent on lumber and wood products. As a result, Oregon is frequently cited as one of the four states with the highest unemployment rates during the early 1980s (the others are Washington, another lumbering state, and Michigan and Ohio, automobile-producing states).

The swing in the economy from the peak of the late 1970s to the extreme trough of the 1980–81 period made the government of Oregon, and the state's Department of Human Resources in particular, search for some relief to the multiple problems posed by growing unemployment. The state government was sensitive to the problem both because of its responsibility for the general welfare and because of the impact on government of dwindling revenues caused by unemployment and substantial tax cuts. The Employment Division of the Department of Human Resources is naturally the state government agency most immediately aware of the impact of unemployment. It collects, assembles, and interprets the data that reveal the segments of the economy and the regions of the state hardest hit by layoffs.

Preliminary Steps toward Work Sharing

During this period, the Employment Division became interested in information on California's Work Sharing Unemployment Insurance program and the federal legislation proposed by Representative Patricia Schroeder of Colorado, which encouraged states to adopt similar programs.

The Oregon Employment Division was kept informed of innovations and initiatives in other states by the Interstate Conference of Employment Security Agencies. In addition, the Department of Labor channeled work-sharing information to all states, including Oregon. This information generated interest in work sharing as one possible solution to Oregon's rampant unemployment. The Governor's office asked the Employment Division to undertake a study of work sharing. The research relied heavily on both written information provided by California's Employment Development Department and frequent conversations with the department's technical staff.

In addition to this direct information on California's legislation and its implementation, the U.S. Department of Labor provided regular updates on similar initiatives in other states and Canada. At the time, Arizona was engaged in similar research and ultimately adopted its own law on April 19, 1981. The U.S. Department of Labor was able to provide drafts of the legislation being considered in Arizona and other states, including Hawaii, Illinois, Maryland, New York, and Pennsylvania, since those states had submitted draft language to the department to make sure that it was in conformity with federal standards.

Both the California and Canadian plans, as well as proposals in other states, were all closely reviewed. Oregon had no desire to reinvent the wheel and no reluctance to adopt the ideas and insights of others or benefit from their hindsight. Ultimately, however, Oregonians were to make their own policy decisions on a wide range of issues that arose once a decision had been reached to propose work sharing supplemented by partial unemployment benefits.

The tentative decision to try what is now called short-time compensation (STC) in federal parlance was reached at policy levels that went beyond the Employment Division and followed informal discussions in and between the executive and legislative branches and also in management and labor circles.

Political issues prevented work sharing from being passed by the regular 1981 legislative session. The Governor had forwarded a work-sharing bill to the legislature in December 1980, but that bill was subsequently incorporated into omnibus legislation that included plant-closure restrictions, provided aid to the employees of closed plants, and encouraged economic development. The omnibus bill died in August 1981 and with it any immediate prospects for a work-sharing program.

Later in 1981 there was a quickly developing and broadening consensus that work sharing would assist employers in maintaining their operations during slack times by helping them preserve trained work forces while avoiding the costs and disruptions of the bumping involved in layoffs. Employers would thus be able to respond more quickly to even temporary increases in market demand. At the same time workers would maintain the job security and income stability that protected them and their families from the traumatic impact—financial, psychological, and social—of total unemployment. Through the Association of Oregon Industries and the Job Service Employer Committees, employers expressed a definite interest. Even though the Oregon AFL-CIO took no official position, there were expressions of similar support from both individual union leaders and labor-oriented legislators.

The benefits to society generally and to government specifically were also seen to be of major importance. Society's commitment to its investment in affirmative action employment for women and minorities could be preserved by work sharing. By distributing the negative impact of recession, extreme gyrations in spending patterns could be moderated.

Unemployment hurts the state in several ways, some obvious and others indirect and less obvious. At the same time that revenues are depressed, increased demands are generated for public assistance and other social services. The cost in unemployment benefits is immediate, large, and easily calculated. In addition there are also the domino effects of unemployment that are reflected in increased criminal justice costs, a higher incidence of juvenile delinquency and school-related discipline problems, and increases in divorce and suicide rates. As employers, state and local governments experience not only the direct cost of unemployment benefits for their employees (they reimburse the unemployment fund dollar for dollar for any payments made to their laid-off employees) but also associated costs due to lowered productivity because of lower morale and the inefficiencies brought about by the bumping that comes with layoffs.

Amending the Unemployment Insurance Laws

While consensus grew that the unemployment insurance laws should be amended to permit a partial subsidy of short-time employment as an alternative to the traditional payment to the totally unemployed, an increasing number of questions were raised regarding the specific ways of effectively and efficiently accomplishing this goal.

Oregon, like other states, already had a system of partial unemployment benefits that permits and encourages a minimal amount of continued work before the weekly unemployment benefit is reduced. This provision, however, could not accommodate any significant amount of work sharing, since benefits are rapidly reduced as work time and earnings are increased. For the typical work-sharing situation in which the employer schedules a three- or four-day workweek, no partial benefits were payable under existing law.

To help educate and inform legislators concerning work sharing, the Employment Division conducted studies and evaluated alternatives, first for the unsuccessful 1981 bill and then, as the economy continued to decline throughout the summer and fall of 1981, for a subsequent bill. Increasing numbers of mills and plants closed their doors without notice, and there was growing public anger and concern. State government responded.

By the time the Oregon legislature met for a special session in January 1982, Arizona had adopted STC legislation, the Canadian program was getting started, and there were encouraging preliminary reports on the effectiveness of the program in California. In February 1982 the special session passed an omnibus employment bill that included amendments to the unemployment compensation law enabling the Employment Division to implement what the legislation referred to as a shared-work plan. Originally the new law was to take effect in October 1982, but in response to the pressures of the continually worsening economy, the legislature amended the law in late February to have it take effect in July 1982. Under this law, Oregon implemented a short-time compensation program, which it dubbed "Workshare."

Implementation of Workshare

The Employment Division devoted the months of March, April, and May to developing plans and procedures for the administration of the new work-sharing program and to publicizing the program. Information and application forms were distributed to employers during the month of June. During the same four-month period, computer programs were written to process applications and payments. The legislation itself contained answers to the major policy questions, but many administrative decisions had to be made before the program could be implemented. Employment Division technicians in research and statistics, data processing, and tax and benefit processing worked with program staff in planning and preparation.

During this period the employers who had to use the forms were consulted through the Association of Oregon Industries (AOI), and the concepts and issues were shared with the Advisory Council to the Employment Division, which represents labor, industry, and the public. As they were completed, copies of forms to be used and a description of the process were sent to field offices, even though it had been decided that the involvement of these offices would be minimal. They were already receiving questions from employers in their communities.

During the early months of preparation, work proceeded smoothly and within the predetermined time frames. In June and July, however, there was an explosion of interest from employers, and the press release that had been planned for June was delayed while applications from employers were processed.

The second employer to apply for approval of a shared-work plan had over 1,000 employees. To field test forms and procedures, a team was sent to the work site to process claims. When the field test proved the process sound, the rest of the applications were processed by mail between the central office and the employer/claimant. (The two-part form used to claim benefits is filled in by the worker and verified by the employer.)

By the end of July, 26 employers had gained approval of shared-work plans involving 2,387 workers, and 18 employers and 489 workers were added during the following month. The response was greater and quicker than had been anticipated, and the division established a separate work-sharing unit in the central office. Work-sharing pamphlets went to 1,500 employers with the Employment Division's monthly "Labor Trends" mailing.

Oregon legislation had been kept deliberately simple and minimal to provide flexibility. The process developed was straightforward and without red tape. What issues and policies did the law decide and what questions were left for administrative discretion? The experiences and ideas of other states and countries had been utilized, but Oregon legislators and policymakers were not wed to their solutions. Several of the issues raised and Oregon's resolution of them are examined below. Because Oregon computes experience for tax-rate purposes by using a benefit ratio method instead of reserve ratio, it does not have "negative-balance" employers as such. The equivalent would be a high-ratio employer, one whose ratio of benefits paid (and charged to the employer) to payroll is relatively high. Typically, these are employers who have had a high turnover due to the seasonal nature of their work.

Approximately 15 percent of Oregon employers would have a high enough benefit ratio to be assigned a higher tax rate solely by reason of participation in work sharing. The tax implications of work sharing are fully explained to each employer expressing an interest in the program.

Basically, the law permits a higher tax rate to be charged to an employer whose employees receive benefits under the shared-work plan than would otherwise be chargeable. The operative words in the law are: "The rate shall not be more than 3.0 percentage points higher than the maximum rate of the schedule in effect for the next calendar year." For 1982 the maximum tax rate was 3.8 percent of the first $11,000 of each employee's earnings. Using this as an example, a high-ratio employer utilizing the shared-work program may, as a result, trigger a higher tax rate up to a maximum of 6.8 percent—3 percent above the otherwise scheduled maximum.

Resolving the Issues of Work Sharing

52-Week Employer Limit. An employer is any entity with a tax number and may have many plants, offices, and employment sites but can have only one shared-work plan. That one plan may cover several locations and groups of employees of the employer, and once it has been approved it may be in existence for a maximum of 52 weeks.

Not all plants or groups of employees need participate in the shared-work program, and at any time during the operation of the plan the employer may receive approval to add other groups. An employer who chooses a short plan of two or three months may later request that it be extended for up to 52 weeks. The plan, however, ends for all employees no later than 52 weeks after its approval, regardless of when individual groups of employees may have commenced participation. No subsequent plan can be approved "until 52 weeks after the plan ends or the last payment was made. . . ." The program is intended for short-time, temporary economic problems.

26-Week Benefit Limit. Under the law "shared work benefits shall not be paid for more than 26 weeks. . . ." While an employer's plan may be in effect for up to 52 weeks, no individual employee may collect shared-work benefits for more than 26 weeks. Those drawing regular benefits are not restricted to 26 weeks.

STC benefits collected are counted against the maximum total dollars to which the employee is otherwise entitled. In other words, shared-work benefits reduce the amount of regular unemployment compensation that a worker may subsequently collect in the event of layoff. However, since the reduction is minimal, benefits would remain for the worker who does become unemployed.

Again, the principle involved is that the program is for temporary economic problems and should not continue so long as to encourage workers to remain on jobs that have no future or employers to retain employees whom they cannot reasonably expect to provide with regular employment.

On the other hand, a worker participating in a shared-work program continues to be employed, and his or her employer continues to pay UI taxes so that the work may be generating employment credits that could lead to eligibility for regular unemployment benefits in the following year—a year in which a shared-work plan could not be in effect.

The eligibility standard for an employee affected by a shared-work plan is different from the standard for an employee on regular unemployment compensation. Both groups require 18 weeks of work and earnings of $1,000 during a 12-month base-year period in order to be eligible for UI benefits. Prior to June 1983, however, a shared-work employee also had to have had continuous employment for six months full time, or one year part time, immediately preceding the application by the employer for approval of a shared-work plan. The 1983 session of the Legislature eliminated the requirement that the continuous employment occur *prior* to the employer's application.

Percentage Reduction of Hours. "The individual's normal hours of work must be reduced . . . at least 20 percent but not more than 40 percent. . . ." This legislative language grew out of the assumption that a work-time reduction of less than one day is unlikely, impractical, and of so little impact that the unemployment insurance program should not encourage it. At the other limit, reductions of more than two days per week could result in a situation where the individual worker would find it more advantageous to collect a "partial" payment under regular unemployment compensation than to participate in work sharing.

If an employee works more than the hours contemplated in the approved plan, no benefits are paid. If the reduction provided by the plan, for instance, is 40 percent and during any week an individual's reduction is only 20 percent (because of working four days rather than three), no benefits may be paid. If in any given week an individual's reduction is *more* than that provided in the plan (40 percent rather than 20 percent), only the benefit for the *planned* reduction is paid. However, when an individual's earnings are reduced even further, to less than the regular weekly benefit amount, regular unemployment insurance benefits may be paid. The computer automatically calculates shared-work and regular UI and pays the claimant the higher of the two.

An existing plan can be modified if the change is requested and approved during the week prior to the change. Individuals and groups may be added and deleted and the percentage reduction for them changed. This allows for greater flexibility for the employer in responding to market and production needs while allowing sufficient notice for automation in the centralized computer process.

Availability Requirements. Shared-work participants are exempted from the normal availability requirements as long as their availability to their employer is acceptable. Paid leave, including vacation and sick leave, may be taken without jeopardizing shared-work benefits as long as some work was performed for the week claimed. If a holiday or other paid leave falls within the scheduled workweek, work-sharing benefits will be paid. However, if a holiday or other paid leave increases earnings for a week beyond the planned reduction, no benefits are paid. Overtime is identical.

Moonlighting. For earnings from other employers, or from self-employment, a forgiveness rule of one-third the weekly benefit amount is applied. The rest is deducted from the work-sharing benefit dollar for dollar. For example, someone with a regular weekly benefit of $150 who is reduced 20 percent will receive $30 in work-sharing benefits. Any moonlighting earnings up to $50 will be ignored. Anything over that is deducted from the $30. Oregon decided that it was appropriate that work-sharing benefits be impacted by other income such as pensions and earnings in the same manner as regular benefits, since work-sharing benefits are charged against the employer in the same manner as regular benefits.

Three-Person Minimum Requirements. Oregon, because of many small employers, chose not to require a higher minimum, as does Arizona, or a percentage of the work force, as does California. It was feared, however, that the elimination of any minimum could result in abuse.

Job Search. "Availability for work, failure to actively search for work or refusal to apply for or accept work . . ." are not reasons for denial of benefits as they are with regular unemployment insurance. The individual, however, must work as scheduled for the shared-work employer.

There are a few other issues not dealt with and questions not completely answered by the legislation itself. These include:

Proof and Lack Thereof. Consistent with the general legislative intent to keep the process simple, employers are not required to document or prove that a layoff would occur were it not for the adoption of work sharing. Any proof offered would be impossible to audit except, perhaps, after the fact and with the benefit of hindsight.

Nor does Oregon limit participation by failing firms. To require a firm to "know" or "show" that it will recover flies in the face of business reality. The use of work sharing could enable a firm to survive where it might have failed if confronted by the trauma associated with layoff and the attendant difficulty of making a timely response to market demand. In any event, in the case of firms that ultimately fail, work sharing will provide the workers with longer lead time to search for other jobs, to acquire training in alternative skills, to plan retirement or relocation, or to adjust to a reduced income.

Fringe Benefits. Oregon employers are required to report how they intend to handle fringe benefits while on shared-work plans, but they are not required to maintain full fringes.

Collective Bargaining Agents. The rights of workers covered by collective bargaining agreements are protected. Shared-work plans for them must be approved in writing by the agent before they can be approved. In this manner, many of the issues, such as treatment of fringe benefits, may be negotiated.

Centralized Administration. Partly because of information gained from the California experience, it was decided to centralize the administration of the plan in order to provide a consistency of interpretation of the new concept as well as to maximize efficiency and economy of operation.

Report on First Nine Months of Activity

Preliminary reports, for the first nine months of operation (July 1, 1982 through March 31, 1983), from the Employment Division of Oregon's Department of Human Resources regarding utilization and reaction to the STC program are basically consistent with reports on the California and Arizona experiences.

Oregon has paid out $1,301,768 in STC to 5,139 workers employed by 193 firms (25 of them unionized) ranging in size from three to 1,034 workers with an average of 49. The overwhelming bulk of the STC activity is in manufacturing, which accounts for over 70 percent of the work-force participation. Although one-fourth of the employers (predominantly small firms) reported reducing some fringe benefits, only 6 percent of the workers suffered such fringe reductions. STC benefits have averaged approximately $37 paid per week over a 10-week period. One-fourth of the employers had 30–100 percent of their workers on the STC program; one-fifth had 60–80 percent, and another fifth 40–60 percent. Thirty percent of the employers had STC plans covering up to 40 percent of their work force. As of March 31, 1982, Oregon had 159 employers of 3,556 workers covered by STC.

The Employment Department surveyed participating employers after the first six months of operation and found a high level of satisfaction with the program. Eighty-three percent reported that the program has met expectations, 99 percent approved the concept and 94 percent found the program met their needs. When asked whether their employees like the program, 98 percent said they do and offered comments ranging from "Most of them would rather work full time, but seem to appreciate the fact that this is a good alternative when there is a lack of work for them" to "All too well! Sometimes they would rather not work full hours when they are needed."

Employers commented on the general public-policy aspects of the program and credited STC with helping them keep their businesses operating. They asserted: "For once someone in government was using their head when they dreamed up the program."

More specific responses point to problem areas of concern to employers:

1. "The 'waiting week' is not appropriate for this program."
2. "I wish we could extend the program beyond the 26-week period, as the . . . economy has not rebounded as quickly as we had hoped."
3. "In a service business such as ours, we have no way of knowing when the work loads will increase or decrease on a day-to-day basis. With our people eligible for Work Share, more flexibility as regards hourly changes within the reporting period would be beneficial."
4. "I wish it were possible to allow the people with higher incomes to get more compensation."
5. "The program would be better if it were set up with a flexible percentage, e.g., an employee would work 24 hours one week and 32 the following and 24 again without being penalized—percentage compensation should 'slide' with work available."
6. "Program needs more publicity—no one I told about it had heard about it."
7. "It would be beneficial if we could modify our program with a phone call. We do not always know how much time we need to cut some people back as far as two weeks in advance. It seems to be a little too easy to get disqualified on technicalities."
8. "Would help to understand a little clearer the personal income tax repercussions."
9. "I think some of the regulations should be changed—the six-month limit on employment, for example, and the stipulation that an employee *has to work* part of the week to collect benefits."

Employers' requests for greater flexibility have been met in part by a change in the legislation which defines an "affected employee" as one who worked for the employer six months full time or one year part time without the requirement that this work precede the employer submission of the plans. This permits new hires (sometimes unavoidable as essential workers quit during the 52-week plan year) to be added to the program.

"An employer continues paying five days worth of health benefits and other fringes to workers who may be working four days. But the state likes that, because tax dollars probably would catch those health bills otherwise. It is an acceptable cost," says Mary Keski, employee and community relations manager of Circle AW Products Co., a Portland electrical products manufacturer that recently put two-thirds of the employees at one factory on work sharing. Quoted in a March 14, 1983, *Forbes* magazine story, Keski concludes: "The plus in maintaining your work force far outweighs the program's minuses." The pragmatic, positive, but unenthusiastic view of Nellie Fox, political director for the Oregon AFL-CIO was: " . . . we found it an acceptable solution to a bad situation." Libby Leonard, Oregon's deputy administrator for employment, *Forbes* reports, sums up the program: "This is a state program without any federal red tape that works well for everyone."

8.
Canada's STC: A Comparison with the California Version

Frank Reid
Noah M. Meltz

It's the runaway government success story of 1982.
Toronto Star, March 27, 1982

Introduction

In January 1982 the government of Canada announced a work-sharing program in which $10 million from the unemployment insurance (UI) fund, which is a federal responsibility in Canada, was allocated to provide short-time compensation to employees whose hours of work were reduced to avoid layoffs. The response to the work-sharing program far exceeded expectations. By the end of its first year of operation, work-sharing applications had been approved for 8,780 firms involving over 200,000 employees, and the UI funds allocated to the program for 1982 had been increased nineteen-fold to $190 million. In 1983, participation in the program continued to accelerate, with almost 3,700 applications approved in the first three months of the year.[1] The cabinet minister responsible for the program, the Minister of Employment and Immigration, has called the program "a roaring success."[2]

To put these figures in perspective, the work-sharing program in California, which has a labor force approximately the same size as Canada, involved 714 firms and 16,000 employees in fiscal 1980.[3] Thus, in its first year of operation, the Canadian program involved over 12 times as many firms and employees as the California program did during its second year of operation.

The purpose of this chapter is to describe the Canadian work-sharing program and to compare its design and its effects with the American programs, specifically the California program, since it has been the subject of a comprehensive evaluation.[4]

106

Program Design

The design of the work-sharing program introduced by Employment and Immigration Canada (EIC) in January 1982 is very similar to the experimental pilot program implemented in 24 firms in Canada during the 1977–79 period. Evaluations were undertaken of both the pilot program[5] and the economy-wide program.[6]

In order to be eligible for participation in the work-sharing program, the firm must have been in business for at least two years, the magnitude of the work reduction over the life of the agreement must be at least 10 percent and not more than 60 percent, and the expected duration of the reduction must be at least six weeks and not more than 26 weeks (although extensions up to 38 weeks were subsequently approved). A firm is not eligible to participate in the work-sharing program if its work reduction is seasonal, if the firm is involved in a labor dispute, or if the firm is already engaged in a private work-sharing scheme.

The restriction on work reductions that are expected to be temporary (i.e., not more than 26 weeks) differs from the California program in which both temporary and permanent work reductions are eligible. However, this aspect of the Canadian program is difficult to enforce rigorously because of the subjective element in the forecast of when demand for the firm's output will return to normal.

In order to receive work-sharing UI benefits, employees must qualify under the usual UI criteria. If the employee is laid off following the termination of the work-sharing agreement, benefits drawn under work sharing do not affect the amount of conventional UI benefits that may be received. This differs from the California program in which conventional UI benefits are reduced by the dollar value of UI benefits that the employee has drawn while work sharing.

In Canada, under conventional UI, an employee is not eligible for benefits for the first two weeks of the layoff. Under the work-sharing program, the two-week waiting period is deferred to the end of the work-sharing program; i.e., it is eliminated if there is no layoff following the termination of work sharing. The California program, by comparison, imposes a one-week waiting period before benefits can be received.

In California, negative-reserve employers (i.e., employers for which UI benefits withdrawn exceed contributions) are discouraged from participating in work sharing by a requirement to pay an additional UI tax of up to 3 percent of earnings subject to contribution.[7] This issue did not arise in the design of the Canadian program because of the absence of experience-rating of UI contributions in Canada.

In both Canada and California, if the employees are unionized, the firm must obtain the agreement of the union to implement work sharing.

In California, " . . . in cases where no union exists, the employer is free to decide whether or not to participate."[8] In Canada, on the other hand, even in nonunion situations, "evidence must be provided that the management and the employees of the work unit have agreed to work sharing and are jointly requesting work-sharing benefits."[9] Employees who agree to participate in work sharing sign a special attachment to the work-sharing application that indicates their agreement to participate and their choice of an employee to act as their representative in matters concerning the work-sharing agreement.

Management's Reaction

The overall reaction of employers to the use of work sharing was highly favorable in both Canada and California. Table 8.1 presents results for the same nine-point rating scale used in both program evaluations. In Canada, 81 percent of employers reported some degree of satisfaction with work sharing, and only 14 percent reported some degree of dissatisfaction. In California, the results were even more positive, with 86 percent reporting satisfaction and 7 percent dissatisfaction.

Results of the Canadian work-sharing evaluation indicate that, from a management viewpoint, work sharing costs slightly less than layoffs, and this is probably the main explanation for management's acceptance of it. Calculations of the effect of work sharing on labor costs are based on the results of two surveys, an "employer survey" of 296 firms[10] and a "productivity and hiring/training cost survey" of 104 firms.[11]

Table 8.1. Employer Reaction
to Work Sharing

	Canada (%)	California (%)
Extremely Dissatisfied	2.4	1.0
Highly Dissatisfied	2.0	1.4
Dissatisfied	6.9	2.4
Slightly Dissatisfied	3.1	2.4
Neutral	5.2	6.6
Slightly Satisfied	7.6	4.5
Satisfied	35.1	31.1
Highly Satisfied	23.0	29.8
Extremely Satisfied	14.8	20.8

Note: The Canadian survey relies on 296 respondent firms, and the California survey involved 292 respondents.
Source: EIC, 1983a, p. 38; State of California, 1982, p. 6.33.

From the employer survey it was found that the average work reduction is 34 percent; work sharing lasted an average of 21 weeks; 90 percent of employees who would have been laid off would have been available for recall; and 28 percent of the work-sharing employees were laid off at the termination of the work-sharing program.[12] It was also estimated that the average weekly wage of sharers was $359 per week, and the average wage of the more junior employees designated for layoff was $322 per week.[13]

From the productivity and hiring/training costs survey it is estimated that the average severance costs of a layoff are $70 per employee (composed of $60 of benefit costs which cannot easily be terminated until the end of the month and $10 administrative costs). To recall a laid-off employee costs on average $325, primarily reflecting the effect of reduced productivity until the employee regains his or her former production speed. The average cost of hiring and training a new employee is estimated to be $944, composed of $179 hiring costs, $90 off-the-job training, $350 on-the-job training, and $325 as a result of below-normal productivity on the job.[14] Table 8.2 summarizes the calculations of the average effect on labor costs of using work sharing rather than layoffs, expressed as a percentage of total wage costs under layoffs.

From table 8.2 it can be seen that the typical Canadian work-sharing firm saved about 1.2 percent of its wage costs by avoiding the rise in

Table 8.2. The Impact on Labor Costs of
Using Work Sharing Rather Than Layoffs

Item	Percent Change in Costs	
	Canada	California
Wages	−1.2	−2.1
Fringe Benefits	+0.8	+2.3
Payroll Taxes	+0.4	−0.3
Total Payroll	0	−0.1
Severance	−0.1	—
Recall	−0.5	−0.6
Hiring and Training	−0.1	−7.0
UI Administration	+0.2	—
Subtotal	−0.5	−7.6
Total	−0.5	−7.7

Note: The difference in cost between work sharing and layoffs is expressed as a percentage of the wage costs under layoffs.

Source: Calculations by the authors from data given in EIC 1983a, p. 120 and State of California 1982, p. 1.9.

the average wage that would have occurred under layoffs when the lower-wage junior employees were laid off. This saving was just offset by the increase that work sharing caused in fringe benefit costs and payroll taxes (Canada Pension Plan contributions and UI premiums).

Severance and recall costs are lower under work sharing than layoffs by 0.1 and 0.5 percentage points respectively, according to table 8.2, because fewer employees are laid off and subsequently recalled. Hiring and training costs are also lower (by 0.1 percent) because the work force is maintained intact under work sharing, eliminating the need to replace those workers who do not respond to recall. Although hiring and training costs are a substantial amount per employee replaced ($944), their effect is of small empirical magnitude because of the high percentage of laid-off workers assumed to be available for recall (90 percent). Overall, the use of work sharing rather than layoffs results in a cost saving equal to about 0.5 percent of wage costs.

Table 8.2 also shows the effect on costs of using work sharing rather than layoffs in California. The smaller impact of work sharing on fringe-benefit costs in Canada than California very likely reflects, in part, the lower costs of medical insurance in Canada due to the subsidy of hospital and medical insurance costs from general government revenues.

The most important single difference in costs between California and Canada is the substantially larger estimated savings on hiring and training costs in California. This reflects two factors. First, the cost of hiring and training a new employee was estimated to be $3,023 in California, substantially higher than the $944 estimated in Canada. Second, the percentage of employees available for recall was assumed to be lower (75 percent in California compared to 90 percent in Canada). This implies a need to replace 25 percent rather than 10 percent of laid-off employees.

An attempt was also made to assess the impact of work sharing on productivity. Management at 53 percent of the 104 firms in the productivity and hiring/training costs survey felt that productivity was higher under work sharing than it would have been under the layoff alternative. Productivity was felt to be lower than under layoffs at 19 percent of the firms.[15] Caution must be used in interpreting these estimates, however, since they involve comparisons with a hypothetical layoff alternative.

Actual empirical data on productivity were available at only 12 firms. For these 12 firms the trend of productivity growth was examined before and after the implementation of work sharing. Results indicated that productivity increased in 7 out of 12 cases. On the other hand, on the basis of *opinions* expressed by managers in the full sample of firms, slightly more felt productivity had decreased (25 percent) than increased (19 percent) as compared with the period prior to work sharing. As the authors of the survey report point out, however, these estimates are of dubious validity, since they do not adequately control for other influences on productivity growth, particularly changing economic conditions.[16] On the basis of this evidence, the use of work sharing rather than layoffs has little or no discernible impact on productivity.

Although the individual firms involved in work sharing in Canada have reacted very favorably to the program, several national business associations have opposed it. The main fears of the central management associations were based on the belief that work sharing leads to under-utilization of workers and presumably lower productivity.

One newspaper reported: "Mr. Roy Phillips, President of the Canadian Manufacturers Association, said that trying to bribe industry to keep people employed, when they can't be employed effectively, isn't helpful to anyone in the long run."[17] Phillips felt that it would be better to use the funds for encouraging investment in manufacturing. Stanley Roberts, president of the Canadian Chamber of Commerce, said that it is improper to use UI funds for any purpose other than unemployment insurance: "Are employees and employers going to have to pay substantially higher rates because UI has become a welfare rather than an insurance program."[18]

The Canadian Federation of Independent Business in a 1982 survey found that its member employers were more than two to one against " . . . the government subsidizing worksharing plans by means of un-employment insurance benefits."[19]

More recently there appears to be more moderation in the central business reaction. For example, Peter Doyle, industrial relations director for the Canadian Manufacturers Association concluded:

> I don't think we can really condemn it [worksharing]. It is obviously keeping people employed. It seems to be the sort of co-operative approach between companies and employees that you can look for in order to improve cost situations.[20]

The magnitude of the response to the work-sharing program in 1982 indicates that this viewpoint is shared by an increasing number of managers.

Labor's Reaction

The reaction of employees involved in the Canadian work-sharing program has been overwhelmingly favorable. A telephone survey of 392 employees involved in the program indicated that 94 percent would support the implementation of work sharing again if a similar situation occurred in the future.[21] A mail survey of 20,906 employees who participated in the program indicated that 89 percent would be willing to participate in work sharing again. A similar level of support was revealed in a survey of 454 employees in the California work-sharing program, which showed 90 percent in favor of repeated use of the program.

At first glance such overwhelming support may seem inconsistent with the fact that the majority of the employees are senior employees who would *not* have been subject to layoff and are experiencing a decline in income by participating in the program. The key to understanding the favorable reaction of senior workers is that they are receiving a significant increase in leisure time in exchange for a relatively small reduction in weekly income.

In a previous work[22] we have demonstrated that if the length of the standard workweek reflects the preferences of the average employee, the average employee will prefer to engage in some amount of work sharing.[23] In addition to the increase in leisure time, the senior employees are able to prevent the layoff of some of their fellow employees.

More specifically, in the Canadian program the average work reduction was 34 percent of weekly hours (i.e., 1.7 days per week). For an employee with earnings below the UI ceiling level ($385 per week in 1983), the employee would receive earnings equal to 66 percent of normal weekly earnings for the days he or she works, plus work-sharing UI benefits of $.60 \times 34$ percent = 20.4 percent of normal weekly earnings, for total income of 86.4 percent of normal weekly earnings. Because some employees had earnings above the ceiling level, the average work-sharing employee received slightly less than this proportion, i.e., income of 84 percent of normal weekly earnings.[24]

For those employees who would have been laid off, work sharing has psychological and social benefits as well as economic ones. Since layoffs are avoided (or postponed), so is the more severe drop in income. In addition, the stress and loss of self-esteem often associated with layoff and full-time unemployment are greatly reduced. These conclusions were supported by the results of the Canadian employee telephone survey, in which 24 percent of the respondents indicated that they would have suffered emotional, health, or marital problems if they had been laid off.[25]

Since work sharing has been shown to have positive economic and social effects on workers, unions would be expected to support its introduction and expansion. While local unions in Canada have, in fact, firmly supported work-sharing programs, central labor organizations vigorously opposed them until late 1982. The reasons for this split lie partly in the differences in time horizons and in the proximity to the workers threatened by layoff. In addition, a special factor in Canada is the difference in the position and role of trade unions compared with unions in the United States. Each of these aspects will be considered.

The fact that local unions are more directly affected by layoffs than the central labor organizations is likely to lead them to place a higher value on programs that alleviate or even postpone layoffs. In addition, central federations may fear the implications of a program that disregards the sacrosanct union principle of seniority. Under work sharing all workers are treated the same regardless of their length of service with an employer. Unions fear that this could be the thin edge of the wedge to remove the application of seniority provisions from other areas of the contract. Central labor organizations also fear that work sharing could be used to force workers into further concessions in other areas.

The difference in attitude toward work sharing on the part of central federations could also be affected by the role and factors relating to the position of the labor movement in Canada. At the beginning of the 1980s approximately 40 percent of nonfarm workers were union members in Canada,[26] almost double the approximately 25 percent rate in the United States.[27] The difference in the rates of organization is even more dramatic when it is realized that in the early 1960s the membership rates were almost identical at approximately 30 percent.[28]

The widening gap between the union membership rates in the two countries is the result of two factors: (1) the rapid organizing of public-sector employees in Canada, facilitated by the passage of legislation permitting workers to organize and in most jurisdictions to strike; and (2) provisions of legislation that provide for certification of unions without representation votes (if the union has a majority of the bargaining unit as members). Labor relations boards have also vigorously and quickly enforced provisions outlawing unfair labor practices.[29]

A major factor in the more prounion (or more genuinely neutral) climate of labor relations in Canada would seem to be the existence of the New Democratic Party (NDP), a social democratic party that receives about 20 percent of the vote federally, and has formed the government in three provinces in Western Canada, and presently is the governing party in Manitoba. The NDP was established in 1961 through a coalition of the labor movement and the Co-operative Commonwealth Federation (CCF), the former socialist party.

The result of the labor movement-NDP alliance has certainly been to focus the labor movement on broad economic as well as social issues. Work sharing deals with both. The central labor federations feared that work sharing, by lowering the measured rate of unemployment, would reduce the pressure on the government to stimulate the economy and attain full employment. Reflecting this concern, unions have referred to work sharing as "poverty sharing."[30] This view is consistent with the central federations' involvement in the political process, even though it conflicts with their role in promoting social justice through the alleviation of the effects of unemployment.

The deepening of the recession and its prolongation appears to have led to a modification of the views of the central federations. The first stage was the realization that work sharing is " . . . incredibly popular with the members and particularly helpful to those who would have been laid off."[31] The second stage seems to have been a recognition of the reluctance of the federal government to adopt strongly expansionary monetary and fiscal policy even in the face of sustained unemployment rates in excess of 12 percent. As a result, the federal Minister of Labour has reported that "labour groups in the pre-budget consultations argued for . . . a continuation of the worksharing program."[32]

Government's Reaction

The Canadian government's interest in work sharing has risen and fallen with changes in the unemployment rate, as the authors have documented elsewhere.[33] In August 1977, following a steady rise in the unemployment rate to 8.3 percent from 7.0 percent two years earlier, the Unemployment Insurance Act was amended to permit work sharing on an experimental basis. Pilot projects at 24 firms were implemented during the period 1977 to 1979.

In a 1980 government report evaluating the pilot projects,[34] it was found that work sharing was very well received by both the firms and the employees in the program, but due to a number of special features of the pilot program, work-sharing UI benefits cost the government 2.4 times as much as conventional UI benefits would have cost. The special features accounting for this cost increase were the absence of a waiting period, a more generous benefit formula at some firms, and the fact that an employee's eligibility for conventional UI was unaffected if a layoff occurred at the termination of the work-sharing program. Although the report noted that work sharing would cost no more than conventional UI if these special features were eliminated, it nevertheless recommended rejection of work sharing. At the time the report was released, the unemployment rate had declined to 7.3 percent.

In December 1981, following a rise in the unemployment rate to over 8 percent, the government allocated $10 million for an economy-wide work-sharing program with the same features as the pilot program (except for the more generous benefit formula). In 1983, as the unemployment rate climbed to over 12 percent, funding for the program increased to $190 million, and participation increased rapidly.

In a recent paper, Leslie Pal of the Department of Political Science at the University of Calgary has severely criticized the process of government policymaking reflected in the above sequence of events.[35] His view is that the Canadian government ignored the conclusion of its own evaluation of the pilot project and introduced the program as an ad hoc response to the crisis of rising unemployment, instead of implementing policy on the basis of careful long-run planning. Pal also suggests that since the UI account is separate from the public accounts, the use of UI funds for work sharing and job creation may have been a way of avoiding the constraints on direct government spending that were implemented at the time.[36]

Ideally, government policy decisions should reflect overall benefits and costs to society, including distributional goals. In both the Canadian and California evaluations an attempt was made to consolidate the costs and benefits to three groups affected (labor, management, and government) to obtain an overall benefit-cost ratio for society. In some ways the overall evaluation for society as a whole is easier than the evaluation for a particular group because many of the effects net out when aggregating over all three groups. For example, an increase in expenditure on fringe benefits is a cost to management but a benefit to labor and thus cancels out in the calculations for society as a whole. A comprehensive and useful tabulation of the theoretical costs and benefits to the various groups and to society as a whole is given in a paper by Juan Mesa.[37] His model shows that for society as a whole, the effect of work sharing is the net impact on recall and training costs, administrative costs, social expenditures related to unemployment, and the effect on productivity.

The EIC evaluation of work sharing specifically recognizes these four categories of effects in its overall benefit cost calculation, but is able to measure satisfactorily only two of them. A typical work-sharing situation involving 50 employees is estimated to reduce the cost of layoff, recall, hiring, and training by $5,555 per year compared to the layoff alternative. Work sharing also involves extra administrative costs for government and the firm of $4,233 per year. This yields an overall benefit-cost ratio of 1.3.[38]

Although this estimated benefit-cost ratio is very close to the benefit-cost ratio of 1.4 estimated in the California evaluation, there are some important differences. The California evaluation includes a reduction of income tax revenues as a net cost to society, but according to the principle outlined in Mesa's paper, this effect is a benefit to workers and, therefore, nets out for society as a whole. It has been estimated that making this adjustment to the California calculations would raise the benefit-cost ratio from 1.4 to 3.2.[39]

More importantly, because of measurement difficulties, neither evaluation includes an estimate of the reduction in the cost to society of dealing with behavior related to the stress of layoffs, for example, alcoholism, mental illness, and crime.[40] As an illustration of the potential magnitude of this effect, Reid has estimated (on the basis of Brenner's work) that the cost to society of dealing with the social consequences of unemployment may equal 25 percent of the direct UI costs.[41] In the work-sharing calculations in the Canadian evaluation, this factor would amount to about $18,742 per year,[42] resulting in the total benefits of work sharing rising to $24,297 and the benefit-cost ratio increasing to 5.7.

In addition, these benefit-cost calculations do not include the value to the workers themselves of the reduction in stress due to the avoidance of layoffs, nor do they place a value on the more equitable distribution of income and leisure that characterizes work sharing.

Conclusions

In this concluding section some speculation is offered on the reasons for the significantly greater degree of participation in the Canadian work-sharing program than in the California program. The main conclusions of the chapter are then briefly summarized.

Three factors that may have contributed to the higher rate of utilization of work sharing in Canada than in California are the higher level of unemployment when the program was introduced, the lower costs of participating in work sharing in Canada, and the greater degree of publicity given to the program.

A primary reason for the high level of interest in work sharing in Canada was that it was introduced in 1982 at a time when the Canadian economy was entering a severe recession and unemployment had risen dramatically. In California, on the other hand, the program was introduced in 1978, at a time when the American unemployment rate was declining from a peak in 1975.

The acceptance of Canadian work sharing was also increased by the fact that the costs of participating in work sharing are less for employers there, since in Canada there is no experience rating of unemployment insurance, and Social Security and medical insurance costs are lower because of greater public subsidies. The costs of participating are also lower for workers in Canada because eligibility for regular UI is not affected in the event of a post-work-sharing layoff.

Utilization of the program has also been encouraged by the tremendous amount of media coverage that work sharing has received, including innumerable newspaper articles, radio and television interviews, and discussions on open-line programs. At first the media tended to view work sharing negatively,[43] but within six months the view had become more positive.[44] Did the public discussion of work sharing in the early phase, even though negative, tend to draw attention to the program?

We cannot say without more in-depth research. It is clear, however, that the change in the media view by mid-1982 was a factor in the further promotion of work sharing. It had been judged a success, and it was highly publicized.

Another major factor in the high visibility of the program was the personal identification of Lloyd Axworthy, the Minister of Employment and Immigration, with the program. The Minister made the work-sharing program a key element in his department's attack on the high and ever-growing unemployment rate and discussed it at every opportunity.

By the end of 1982, UI-funded work sharing had become a widely used and popular labor-market program in Canada. Many writers (the authors included) would prefer a more expansionary monetary and fiscal policy that would reduce the need for either layoffs or work sharing. When faced with the alternative of layoffs, however, work sharing has proved to be a better alternative for both management and the employees in the thousands of firms where it has been used. The initial rejection of work sharing by the central labor organizations has been replaced by active lobbying for the program's continuation, and the resistance of the central employer federations has also moderated. In addition, Employment and Immigration Canada, the government department that administers the work-sharing program, has reversed its assessment from a negative one in its 1980 evaluation of the pilot program to a positive one in the 1983 evaluation of the economy-wide program. Although some modifications to the program are likely on the basis of this evaluation, it is safe to say that work sharing has now become an integral part of labor-market policy in Canada.

NOTES

1. Employment and Immigration Canada, *A Preliminary Evaluation of the Work Sharing Program* (Ottawa, Ont.: Employment and Immigration Canada, 1983a), p. 16.

2. Lloyd Axworthy, Canadian Minister of Employment and Immigration, quoted in the *Toronto Star*, 27 March 1982, p. A4.

3. State of California Health and Welfare Agency, *California Shared Work Unemployment Insurance Evaluation* (Sacramento, Calif.: State of California Health and Welfare Agency, 1982), p. ES 1.

4. Ibid.

5. Employment and Immigration Canada, *Evaluation of UI-Funded Work Sharing in Canada* (Ottawa, Ont.: Employment and Immigration Canada, 1980).

6. Employment and Immigration Canada, *A Preliminary Evaluation of the Work Sharing Program*.

7. State of California Health and Welfare Agency, *California Shared Work Unemployment Insurance Evaluation*, p. 3.4.

8. Ibid., p. 3.4.

9. Employment and Immigration Canada, *A Pre-Evaluation Assessment of the UI-Funded Work Sharing Program* (Ottawa, Ont.: Employment and Immigration Canada, 1982), p. 133.

✳10. CanWest Survey Research Corporation, *The UI-Funded Work Sharing Program: The Experience of Employers and the Role of Work Sharing in Preventing Layoffs* (Winnipeg, Man.: CanWest Survey Research Corporation, 1983).

✳11. DPA Consulting, Ltd., *CEIC Work Sharing Program: Analysis of Productivity Changes and Hiring and Training Costs* (Ottawa, Ont.: DPA Consulting, Ltd., 1983).

12. The high percentage of laid-off employees available for recall (90 percent) is a result of the exceptionally high level of unemployment in 1982. Firms indicated that on average during the 1970s the percentage of laid-off employees who responded to recall was 45 percent.

13. Employment and Immigration Canada, *A Preliminary Evaluation of the Work Sharing Program*, pp. 114–117.

14. DPA Consulting, Ltd., *CEIC Work Sharing Program: Analysis of Productivity Changes and Hiring and Training Costs*, p. 68.

15. Ibid., p. 28.

16. Ibid., p. 22.

17. Wilfred List, "Work Sharing Plan Gets Cool Reception," *Globe and Mail*, 12 December 1981.

18. Robert Stephens, "Diverting UI Money for Work Sharing Assailed as Welfare," *Globe and Mail*, 30 May 1982.

19. Canadian Federation of Independent Business, *Mandate No. 93* (Willowdale, Ont.: Canadian Federation of Independent Business, 1982), quoted in Employment and Immigration Canada, *A Preliminary Evaluation of the Work Sharing Program*, p. 83.

✳ 20. Virginia Galt, "Few Firms Expected to Seek Union Concessions," *Globe and Mail*, 8 June 1982.

✳21. Employment and Immigration Canada, *Attitudes and Experiences of Participants in the Work Sharing Program* (Ottawa, Ont.: Employment and Immigration Canada, 1983b), p. 107.

22. Noah M. Meltz, Frank Reid, and Gerald Swartz, *Sharing the Work: An Analysis of the Issues in Worksharing and Jobsharing* (Toronto, Ont.: University of Toronto Press, 1981).

23. In economic terms, work sharing results in a kink in the budget line in the income-leisure choice diagram. If working hours prior to work sharing reflect an initial equilibrium position, then the new equilibrium will be on a higher indifference curve with decreased income, increased leisure, and a higher level of satisfaction.

24. Employment and Immigration Canada, *A Preliminary Evaluation of the Work Sharing Program*, p. 147.

✳ 25. Employment and Immigration Canada, *Attitudes and Experiences of Participants in the Work Sharing Program*, pp. 97–98.

26. Labour Canada, *Labour Organizations in Canada* (Hull, Quebec: Ministry of Supply and Services Canada, 1982).

27. Courtney D. Gifford, *Directory of U.S. Labor Organizations, 1982–83 Edition* (Washington, D.C.: Bureau of National Affairs, 1982).

28.a. Labour Canada, *Labour Organizations in Canada*.

b. Thomas A. Kochan, *Collective Bargaining and Industrial Relations: From Theory to Policy to Practice* (Georgetown, Ont.: Irwin-Dorsey, 1980).

29. Alastair Hugh Halliday, *The Effects of Legislation on Trade Unions in Canada and the United States* (Cambridge, Mass.: Department of Economics, Harvard University, 1982).

30. Virginia Galt, "Threat of Layoffs Makes Work-Sharing More Popular," *Globe and Mail*, 29 March 1982.

31. Wilfred List, "Pay Cuts, 4-Day Week: Work-Sharing Helps Avoid Layoffs," *Globe and Mail*, 27 May 1982, quoting Ray Hainsworth, education director of the Ontario Federation of Labour.

32. Charles Caccia (Minister of Labour), *The Fourth Day of Debate on the Budget: Notes for an Address* (Ottawa: Labour Canada, 1983).

33. Noah M. Meltz and Frank Reid, "Reducing the Impact of Unemployment through Worksharing: Some Industrial Relations Considerations," *Journal of Industrial Relations* 25 (June 1983).

×34. Employment and Immigration Canada, *Evaluation of UI-Funded Work Sharing in Canada.*

35. Leslie A. Pal, "The Fall and Rise of Developmental Uses of UI Funds," *Canadian Public Policy*, vol. 9, no. 1, pp. 81–93.

36. Ibid, p. 89.

37. Juan M. Mesa, *Short-Time Working as an Alternative to Layoffs: The Case of Canada and California*, ILO Working Paper (Geneva, Switzerland: International Labour Organisation, 1982), table 1.

38. Employment and Immigration Canada, *A Preliminary Evaluation of the Work Sharing Program*, p. 132.

39. Frank Reid, *Conceptual Issues in the Evaluation of Work Sharing in Canada* (Ottawa, Ont.: Employment and Immigration Canada, 1983), p. 29.

40. Harvey Brenner, *Estimating the Social Costs of National Economic Policy: Implications for Mental and Physical Health and Criminal Aggression*, Joint Economic Committee, U.S. Congress (Washington, D.C.: Government Printing Office, 1976).

41. Frank Reid, *Conceptual Issues in the Evaluation of Work Sharing in Canada*, p. 20.

42. Assuming 50 work sharers draw UI 1.7 days per week at $42 per day for 21 weeks, this implies total UI costs of $74,970 per year. Estimating the social costs of the unemployment that would have occurred if layoffs had been used at .25 percent of this amount implies a cost savings of $18,742 per year due to work sharing.

43. Wilfred List, "Work-Sharing Plan Gets Cool Reception," *Globe and Mail*, 12 December 1981.

44. a. Carol Goar, "Work-Sharing: A Success No One Really Expected," *Toronto Star*, 27 March 1982.

b. Wilfred List, "Pay Cuts, 4-Day Week: Work-Sharing Helps Avoid Layoffs," *Globe and Mail*, 27 May 1982.

9.
The Federal Response to Short-Time Compensation

Linda A. Ittner

When a federal bill on short-time compensation was signed into law on September 3, 1982, it brought to fruition the efforts of a broad-based coalition that had labored since the mid-seventies to make a new and more acceptable form of work sharing available to Americans as an alternative to layoff.

Work sharing is not a new concept. Indeed, until recently it has been considered an obsolete one which has been supplanted by other unemployment strategies. During the Great Depression, President Herbert Hoover relied heavily on voluntary work sharing as a way to reduce unemployment without fiscal or monetary federal intervention. Under Hoover's work-sharing plan, employers frequently cut workers' hours as well as wages in the absence of a federal minimum wage and hours law or union contracts in most industries. And, of course, there was no unemployment compensation system at that time for those totally without work.

Organized labor has bitter memories of work sharing as practiced during the Hoover years, since it forced workers to "share the misery" of unemployment with no cushion against joblessness. When unemployment rose above 20 percent during the Depression, work sharing was one of the only tools available to reduce unemployment. Both employers and their employees recognized that total layoffs of part of the work force would mean destitution for many of those who were laid off. Workers received few fringe benefits so that fixed costs per employee were low. This encouraged employers to conclude that work sharing would not only be more humanitarian, but also more economical, since older, highly paid employees would have their hours cut back along with the younger, lower-paid workers. Even if a majority of employees preferred layoffs for the few instead of work sharing for the many, they could not prevent the employer from unilaterally instituting work sharing.[1]

Work sharing was also used to some extent during the Roosevelt Administration (see chapter 11 for a more detailed description of this era), but it was only one of several approaches to solving the problem of unemployment. The existence of unemployment insurance after 1938 encouraged a seniority-based system of layoffs rather than work sharing, since it guaranteed a basic income for lower-seniority workers who were laid off. Thereafter, work sharing, as it was known during the Depression, gradually fell into disuse.

More frequent recessions in the seventies—most notably, the major recession of 1974–1975—and rising unemployment fueled a renewed interest in work sharing as an alternative to layoffs. But this was a new kind of work sharing, one that permitted the partial payment of unemployment benefits for days of work lost. It was called short-time compensation (STC) and had been in use in Germany since the 1920s.

In 1974, Martin Nemirow, a policy analyst at the Department of Labor, wrote a paper setting forth in broad outline the basic idea for STC. It was part of a policy-planning exercise to deal with the onset of the recession that lasted through 1975, but no action was taken on the proposal at that time.

At the same time, civil-rights activists saw the devastating effect that layoffs were having on minority workers and women. Title VII of the Civil Rights Act of 1964, which prohibited employment discrimination on grounds of race, color, religion, sex, and national origin, became effective in 1965. By the early 1970s, employers had begun to hire and promote minorities and women into jobs that they had not traditionally held. When the recession came along, these newly hired employees were the most susceptible to layoff because of low seniority.[2]

In April 1975, the New York City Commission on Human Rights, chaired by Eleanor Holmes Norton, sponsored a two-day conference to explore alternatives to layoffs. Representatives of labor, management, and academia surveyed a broad array of alternatives to layoffs, which included: mandatory cuts in executive and middle management salaries, reduction in fringe benefits, voluntary furloughs, early retirement, and work sharing.

The New York conference focused on the experience of six companies that had instituted work-sharing programs. These cases indicated that work sharing without government incentives usually arose in atypical circumstances—when the firm's existence was threatened and labor-management relations were unusually good.

The conference concluded that work sharing in the form of "a shorter work week, or rotating and staggered shifts, or any other method by which average work hours are reduced" was the layoff alternative with the widest potential application to recession-based economic problems and almost all types of business and industry.[3] These discussions also indicated a strong interest in work sharing; indeed there was more receptivity than might have been anticipated, especially on the part of unions.[4] The conference emphasized that work sharing was "not a panacea" and recommended that any more than a 20 percent work reduction would create too much hardship for the work sharer. The conference also pointed to the need to offer European-type incentives to stimulate more work sharing.

On September 5, 1975, a 15-member Task Force on Unemployment appointed by New York Governor Hugh Carey released a work-sharing proposal recommending that the state unemployment law be amended to permit payment of unemployment benefits to workers laid off for one day a week. The unemployment insurance (UI) system discouraged work sharing because partial benefits were limited to the difference between full weekly benefits and any income earned in the week. As a result of this restrictive interpretation of partial benefits, a worker would have had to work less than half time to qualify for any UI benefits.

This work-sharing-UI proposal, named the Poses Plan after its chief sponsor on the Governor's Task Force, Lillian L. Poses, a lawyer and former regional counsel of the War Manpower Commission, was endorsed in a *New York Times* editorial on September 30, 1975, which cited New York City's 12 percent unemployment rate and urged Governor Carey to recommend a work-sharing amendment to the state UI law.

In response to the recommendation of the Governor's Task Force, Assemblyman Seymour Posner introduced Assembly Bill 8737 on June 26, 1975, to permit UI benefits to be paid to partially unemployed workers. The bill died in committee.

The following year, Peter Henle, a senior policy analyst with the Congressional Research Service, published a report on work sharing that commanded wide attention. In it, he reviewed the widespread use of STC abroad and called for serious consideration of developmental efforts in the United States. His work contributed to President Carter's support of the general concept in the 1976 presidential campaign.

In 1976 the U.S. Civil Rights Commission, disturbed by the impact of the recession on affirmative action efforts in employment, held an informal hearing in Washington to solicit comment on a commission draft report, "Last Hired, First Fired: Layoffs and Civil Rights," which addressed the issues of seniority, layoffs, work sharing, and other civil rights aspects of national economic policy.[5]

The Commission's report made some specific recommendations for alleviating the disproportionate burden of layoffs on the "last hired and first fired," including:

- Revision of state unemployment insurance laws to provide tax-free UI benefits for employees choosing to work a reduced workweek.
- Federal guidelines by the Equal Employment Opportunity Commission (EEOC) that would stipulate that when an employer is compelled to reduce production it should be done in a way that does not adversely impact minorities or women.
- Alternatives, such as reduction of hours, early retirement, rotation of layoffs, cuts in costs other than wages, and other techniques.
- Office of Federal Contract Compliance Programs (OFCCP) guidelines on layoffs and recalls, consistent with Executive Order 11246, which prohibits discrimination in employment by federal government contractors and subcontractors on the basis of race, creed, color, or national origin and requires contractors to take affirmative action to ensure that equal opportunity be provided.[6]

To further explore the feasibility of STC, Dr. Fred Best undertook survey research in 1977 which looked at the attitude of workers, employers, and unions toward STC. In 1979, Dr. Best was appointed as a liaison by the U.S. Department of Labor to coordinate cooperative research on California's STC program. Ultimately, Dr. Best developed an evaluation of the program for the state of California.[7]

California Tries Work Sharing

On November 1, 1977, the California Select Committee on Investment Priorities and Objectives, chaired by Senator John F. Dunlap, held hearings in San Francisco on "Leisure-Sharing: Is There a Future in Spreading the Work?"[8]

Senate President Pro Tem James Mills, who requested the hearing, said that economic growth could no longer meet growing job needs, public-service jobs were a palliative, and reversing technology to make jobs was impossible. He said that Americans must investigate the "fourth strategy," redistributing existing hours of work so more people are hired to produce the same output, thus encouraging job holders to enjoy more leisure time so others may work.

Two other senators expressed interest in legislation to promote work sharing. Chairman Dunlap was considering legislation to provide a tax incentive to employers or employees to defray the cost of fringe benefits and extra costs associated with reducing work hours as a means of creating more jobs. Senator Bill Greene testified about his interest in legislation to change the unemployment insurance law to permit UI payments for partial time loss in connection with work sharing as an alternative to layoffs.

Greene and Robert Rosenberg of the Senate research staff continued to study the use of unemployment insurance to supplement reduced work hours. They met with the people involved in the New York efforts to enact state work-sharing UI legislation and traveled to Europe to investigate European STC policies.

In June 1978, California adopted Proposition 13, a measure that limited property tax assessments and threatened to force massive layoffs in the state's public work force. This development accelerated efforts to establish a state work-sharing program. Within a month, Senator Greene's work-sharing UI legislation was enacted.

The Washington Connection

In August 1977, Eleanor Holmes Norton met with Secretary of Labor Ray Marshall and Assistant Secretary for Employment and Training Ernest G. Green to urge that the Labor Department take action on work sharing.

In November 1977, Green proposed a work-sharing demonstration project to be funded outside the federal-state unemployment insurance system to avoid any legal objections, since at that time no state permitted unemployment benefits for one day of layoff per week. Since STC could not be paid from the UI trust funds, it was suggested that benefits be paid from Comprehensive Employment and Training Act (CETA) discretionary resources. If internal rather than appropriated funds were used, it was anticipated that some preliminary findings should be available within a year or 18 months.[9]

The Green memo noted that AFL-CIO representatives had expressed serious objections to a proposal by EEOC that states implement a work-sharing UI program to be eligible to receive offset credit under the Federal Unemployment Tax Act.

In 1978 a Department of Labor work group was established, including representatives of the Employment and Training Administration (ETA) and the Office of the Assistant Secretary for Policy, Evaluation and Research (ASPER), to develop a work-sharing demonstration project. The group, directed by Peter Henle, deputy assistant secretary for policy, evaluation and research, developed a research proposal, competitively awarded to Mathematica, a Princeton, New Jersey, research firm, to develop STC programs at several sites. If the demonstration projects proved successful, it was decided, the program could be expanded.

Despite extensive planning, the Department of Labor dropped the work-sharing demonstration project in July 1979 because of AFL-CIO objections that it was illegal for the department to use CETA discretionary funds to finance the project.

Even though the work-sharing demonstration project was scuttled, there was a continuing discussion between Martin Nemirow of the Department of Labor and John Zalusky and others associated with the AFL-CIO to work out a mutually acceptable design for an experimental STC model. At the same time, there was dialogue among the AFL-CIO staff, California union leaders, and the staff of the state's Employment Development Department, which was implementing California's work-sharing UI program. These discussions and the study by Martin J. Morand and Donald S. McPherson on union responses to California's Work Sharing Unemployment Insurance Program[10] paved the way for eventual AFL-CIO acceptance of short-time compensation legislation introduced by Democratic Representative Patricia Schroeder of Colorado.

Growing Interest in Short-Time Compensation

In January 1978, the prestigious Committee for Economic Development issued a report entitled *Jobs for the Hard-to-Employ: New Directions for a Public-Private Partnership*. It contained a statement on national policy by its Research and Policy Committee that included some policy options for minimizing unemployment during recessions. One of these recommendations was "active exploration of possible legal and administrative changes to facilitate work sharing as an alternative to cyclical layoffs in cases where such a solution is desired by both management and labor."[11]

The National Commission on Unemployment Compensation also considered work sharing during its deliberations. The commission was mandated by law in 1976 to undertake a thorough and comprehensive review of the unemployment compensation program. In August 1979, when work sharing was under discussion, the commission issued a report entitled "The Shared Work Compensation Form of Worksharing: Background and Issues." In its final report, the commission recommended "continued study and evaluation of various partial benefit provisions, including worksharing proposals." It also urged the U.S. Department of Labor, states, and independent research organizations to participate in such studies and evaluations.[12]

On September 12, 1980, EEOC published a statement on layoffs and equal employment opportunity in the Federal Register.[13] The statement expressed concern about the adverse impact of layoffs on minorities and women and strongly urged employers, labor organizations, and others affected by Title VII to find alternative methods of reducing labor costs which do not have a disproportionately adverse impact on minorities and women. It recommended work sharing as one alternative and called upon governors and state legislators to amend their unemployment compensation laws to allow payment of partial unemployment insurance benefits to workers on reduced workweeks. The commission further urged that if a layoff were found to be unavoidable, it should be based on plant-wide rather than departmental seniority in order to lessen the adverse impact on women and minorities. The commission solicited information and suggestions from employers and unions on any layoff alternatives that were currently in use.

The Drive for Federal Legislation

Representative Schroeder introduced H.R. 7529, the Short-Time Compensation Act, on June 9, 1980. The bill was the outgrowth of her legislative interest in alternative work schedules and hearings held by her subcommittee in 1978.[14] In the 95th Congress two bills were enacted that had been under consideration by the subcommittee. The first, the Federal Employees Part-Time Career Employment Act (P.L. 95-437), promoted permanent part-time employment in the federal government with prorated fringe benefits. The second, the Federal Employee Flexible and Compressed Work Schedules Act (P.L. 95-390), established a three-year flexitime experiment which waived the overtime requirements for those working more than eight hours per day on a union-management approved compressed-work schedule.

After the Department of Labor scuttled its work-sharing UI demonstration project, Rep. Schroeder determined that federal legislation was necessary to assure the department's involvement in assisting state STC initiatives and in evaluating state programs.

The Schroeder bill proposed a voluntary approach to STC for states, employers, and unions. The bill authorized the Secretary of Labor to develop model legislation, provide technical assistance to states, and make grants to states for start-up costs of new STC programs. It also established guidelines modeled after the California experience for state programs. The Department of Labor was authorized to conduct one or more demonstration projects for purposes of evaluating the effectiveness of STC and to make a final report to the Congress and the President with any recommendations that the Secretary considered appropriate.

Shortly after introduction, hearings were held by the Ways and Means Subcommittee on Public Assistance and Unemployment Compensation on several UI bills, including Rep. Schroeder's.[15] Support for the Schroeder legislation came from civil rights advocates—Eleanor Holmes Norton, the National Urban League, Alfred and Ruth Gerber Blumrosen, and the American Jewish Committee.

Motorola, Inc. testified from the employer's perspective that STC would benefit both its employees and the company. Motorola began to look for alternatives to layoff after the 1974–75 recession. Chairman of the Board Robert W. Galvin wrote to the Labor Department in support of its proposed STC demonstration project. Senator Bill Greene, sponsor of the California work-sharing UI legislation, and Dr. Fred Best of the state's Employment Development Department testified in support of H.R. 7529 and described the California program. Therman C. Kaldahl, president-elect of the Interstate Conference of Employment Security Agencies (ICESA), urged passage of the Schroeder bill "because it offers flexibility to States and offers us a unique program to deal more creatively with our labor market."[16]

The Department of Labor, AFL-CIO, and Chamber of Commerce all recommended that passage of federal legislation be delayed until the official evaluation of the California program was available. The subcommittee did not take action on the STC bill, although it did pass legislation to extend unemployment benefits from 39 to 52 weeks. Because of the prolonged 1980 recession, large numbers of laid-off workers had exhausted their UI benefits.

During late 1980 and 1981, extensive discussions were held between the Schroeder staff and AFL-CIO representatives who reiterated concerns about STC and insisted that any legislation address 12 specific issues:

- Control costs so that STC would not drain the UI trust funds.
- Protect newly hired minorities and women who might be laid off before work sharing could be instituted for the more highly skilled workers.
- Maintain fringe benefits under a work-time reduction program.
- Provide parity for senior, more highly paid workers in wage replacement benefits.
- Establish federal guidelines that would be mandatory for the states.
- Require agreement of the bargaining representative before implementation of STC.
- Establish criteria for the Department of Labor evaluation of state programs.
- Change the Department of Labor's statistical practices in counting workers on reduced workweek as employed or unemployed, since the distribution of state and federal funds depends upon the unemployment rate.
- Provide funding for state STC programs and their evaluation.
- Require certification of the employer's need for reducing work time before instituting a short-time compensation program.

- Ensure that an employer would not use STC continuously to prevent it from becoming a permanent subsidy.
- Provide for special settings, such as construction and longshoring, where the employer hires through an employer association that is party to a collective bargaining agreement.

When the 97th Congress resumed on April 2, 1981, Representative Schroeder reintroduced her bill (H.R. 3005), incorporating changes that had been suggested by interested groups.

In May 1981, California's State Senate and its Employment Development Department sponsored a Conference on Work Sharing Unemployment Insurance in San Francisco, cosponsored by a wide spectrum of supporters: the Council of State Governments, the National Governors' Association, employers (Motorola, Inc., Caterpillar Tractor Co., Arizona Association of Industries, California Conference of Employer Associations, and the California Manufacturers Association), unions (District 115 International Association of Machinists, Local 11 International Brotherhood of Electrical Workers, and the California State Council Service Employee's International Union), Governor Hugh Carey of New York, Representative Schroeder, and Representative John F. White, Jr. of the Pennsylvania General Assembly.

Conference participants included union leaders, business executives, and representatives from state government who wanted to learn more about the California work-sharing program. Interest in STC was clearly spreading throughout the country.

Although the labor movement's initial response to the Schroeder STC bill was cool, a major change came in August 1981, when the AFL-CIO Executive Council adopted a resolution supporting the concept of work-time reductions and prorated unemployment insurance benefits provided there were adequate safeguards for employees. The five safeguards cited in the statement were: (1) adequate funding; (2) where workers are represented by a union, agreement by the union before implementation of STC; (3) wage replacement level of at least two-thirds of each worker's lost pay for up to 40 percent of the workweek; (4) full retention of pension, insurance, and other fringe benefits; and (5) protection from discrimination against recently hired workers, especially minorities and women.

Not long after this resolution was adopted, Work in America Institute completed an 18-month-long policy study of new work schedules with a report entitled *New Work Schedules for a Changing Society*. The report strongly endorsed short-time compensation and recommended "the enactment of a short-time compensation law embodying the key provisions of the Schroeder bill: the development of model STC legislation, and grants and technical assistance to states to assist them in developing, enacting, and implementing STC programs."[17]

When the second set of hearings was held in the House of Representatives in May 1982, STC had new support from labor witnesses representing the AFL-CIO, United Auto Workers, and the American Federation of State, County, and Municipal Employees.

The bill was amended once more to accommodate the concerns of interested groups and for political expediency in a Congress that was totally dominated by forces pledged to cut the federal budget during the Reagan Administration. To reduce the costs, the bill was stripped of the demonstration project and monetary grants to the states. The $10 million authorization was dropped, with the assumption that the evaluation could be accomplished under the Department of Labor's research and development budget. Other changes included:

- The specification that no employee could collect STC for more than 26 weeks in a 12-month period.
- Broadening the definition of "qualified employer plan" to apply to employer associations.
- The requirement that the work force in the affected unit could not be reduced during the previous four months by layoffs of more than 10 percent to prevent the total layoff of the unskilled work force before adoption of a work-sharing program.
- The requirement that employer plans be reviewed at least annually by state agencies to assure that they continue to meet the requirements of federal and state law to prevent employer abuse of STC.

Fringe Benefits and Short-Time Compensation

The treatment of fringe benefits was a major concern of the unions, which feared that utilization of STC would diminish a worker's Social Security benefits and reduce retirement income for pensioners. To remedy this problem the AFL-CIO proposed that STC benefits be counted as earnings in calculating Social Security benefits.

Rep. Schroeder asked the Congressional Research Service (CRS) to study the potential effect of STC on Social Security benefits. The report was issued in November 1981.[18] CRS based its calculations on a computerized model, using three different wage histories, and found that participation in STC programs would not substantially reduce Social Security benefits. Depending upon the wage history, the typical worker lost from $3 to $6 per month in the simulation. Counting STC benefits as earnings would reduce the losses by one-half.

The CRS simulations were based on a 23-year period beginning in 1956. This excluded two of the six recession years in which STC was assumed to be paid, which mitigated its effect on old-age benefits. If the computation period had been longer, the effect of short-time compensation might have been greater, according to the report.

> For example, in 1991 the elapsed years will reach a maximum of 40 under the new method [of calculation]. Then there will be 35 computation years. If one assumed under present law that most compensated unemployment occurs in the five years with the lowest wages that are dropped from the calculation, the redistribution of compensated unemployment throughout the 40 years might cause a larger decline in Old Age benefits. Without detailed information on actual age-unemployment patterns under present law and short-time compensation, however, the size of the decline is very difficult to predict."[19]

Rep. Schroeder requested another CRS study of the impact of STC on pension plans.[20] This study concluded that STC should have a negligible effect on benefits received from *defined contribution* plans, whether it be a "money purchase" type or a profit-sharing arrangement. Since contributions are made throughout a worker's participation in the plan and interest and earnings would continue to accrue, a 10 percent lower salary—even if it occurred during the last three years of employment—would only have a slight effect on the benefit that could be purchased with the account balance credited over a worker's entire career.

Practically all collectively bargained, private-sector pension plans are *defined-benefit* plans, which provide a stated dollar benefit for each year of service. Under these plans, an STC program would not directly affect the amount of the pension received by the individual. In most instances the individual would still receive full pension credit for each year of service. Under the Employee Retirement Income Security Act of 1974 (ERISA), the term "year of service" means a 12-month period during which the employee has not less than 1,000 hours of service (or, in the case of the maritime industry, 125 days of service).

About 7 million workers participate in collectively bargained, *multi-employer* plans, which provide a stated dollar benefit for each year of service and are usually funded by employer contributions based on hours of work. STC would reduce the amount of contributions flowing into this type of defined benefit plan. The effect of STC on the financial solvency of multiemployer plans or on future levels of benefits cannot be ascertained without further study, according to CRS.

STC could affect pensions received by workers in *final pay* plans. With wages rising over time, pensions based on the final few years' earnings result in a higher benefit for most workers than if they were based upon career average earnings. The impact of STC on retirement benefits differs among plans and workers, reflecting the great variety of pension plan formulas, wage histories, and other variables. The CRS study included an illustration from a "worst case" scenario:

Assume that a worker participated in the STC program for 26 days in each of the last 3 years of employment. Assume further that the plan provided a benefit to a long service worker of 50 percent of final average pay. Being laid off 26 days a year is the same as ½ day a week, or 4 hours out of 40. Rather than receiving a salary based on 40 hours, the worker would receive a salary based on 36 hours (90 percent of 40). Thus, the employee would receive 90 percent of his expected normal annual salary, and part of the 10 percent deficit would be made up by the STC program. Assuming the previous wage history, the final 5 years' salary under the STC program would be as follows:

Assumed Wage History

Year	Old	New (STC)
1977	$16,171	$16,171
1978	17,140	17,140
1979	18,501	16,651
1980	20,000	18,000
1981	21,740	19,566
5-year average	18,710	17,506
3-year average	20,080	18,072

If the monthly pension were based on the final 5-years' average salary, it would amount to $729 for a worker who participated fully in the STC program, rather than $780 (about 6½ percent less). If the monthly pension were based on the final 3-years' average salary, it would amount to $753, rather than $837 (about 10 percent less). In all cases the individuals would receive Social Security in addition to their pensions. [As noted in Rich Hobbie's November 5, 1981, report "Potential Effect of Short-Time Compensation on Social Security Benefits," STC did not substantially reduce Social Security Old Age Benefits in the specific examples used.] If a final pay plan were integrated with Social Security under the so-called "offset method," a lower Social Security benefit would result in a lower pension offset. Roughly, for each $1 lower Social Security benefit, the employer would pay a 50 cent higher pension.

The potential benefit reductions in final pay plans would be further mitigated by the following factors: (1) Many plans base benefits on the 3 or 5 highest years' earnings in the last 10 years (for example, in the wage history used above, the $17,140 earnings in 1978 would be substituted for the $16,651 in 1979), (2) workers may not be on STC for the maximum 26 days nor necessarily in all of the final three years of employment, and (3) if employees did not receive the assumed pay raises in the final years of employment, earnings in the non-STC years would be higher and could in turn be factored into the pension compensation base.[21]

The final version of the Schroeder STC bill recommended that the "employer continues to provide health benefits and retirement benefits under defined benefit pension plans (as defined in Sec. 3 (35) of ERISA) to employees whose workweek is reduced under such plan as though their workweek had not been reduced." This provision proposes full benefits for employees covered by defined-benefit plans, which include about 69 percent of all pension plan participants. The Department of Labor evaluation should carefully monitor the treatment of fringe benefits by employers to assess to what degree they are maintained for employees on STC and to determine what loss, if any, occurs for workers covered by defined contribution, final pay, and multiemployer pension plans.

Enactment of Federal Legislation

Following the hearings in May 1982 before the Ways and Means Subcommittee, chaired by Representative Harold E. Ford (D. TN), Representative Robert T. Matsui (D. CA) offered the Schroeder STC bill (H.R. 5904) as an amendment to the omnibus tax bill.

It is ironic that during a period of double-digit unemployment the major opponent of STC was the Department of Labor (DOL), which argued that since states could already enact STC laws on their own, there was no need for federal legislation. Moreover, DOL considered it inappropriate for the federal government to encourage the states to undertake compensated work sharing. This was consistent with the Reagan Administration's policy to reduce federal government involvement in national social problems but ignored the potential benefits of STC: increased productivity and a reduction in federal costs associated with unemployment.

The Labor Department argued that the mandated study of the state STC programs would cost from $5 to $8 million and could not be undertaken without additional appropriations. Supporters of the federal legislation felt that DOL's cost figures were inflated, since the massive study of the unemployment insurance program in six states and six cities prepared for the National Commission on Unemployment Compensation in 1978 cost $1 million. Congressional supporters were concerned that none of the department's research budget was targeted on cyclical unemployment, despite the high costs of recessions to both workers and companies.

DOL also opposed the federal legislation on the grounds that administrative costs for STC would be higher than for regular UI claims and would increase costs to the federal government, which pays for the administration of the unemployment compensation program. Proponents of the federal legislation admitted that administrative costs of the California

STC program were 15.8 percent higher for work-sharing claims, but asserted that this was caused by the hasty implementation of the program. The higher costs reflected the usual start-up expenses for a program that was administered by thousands of local UI offices frequently unfamiliar with the new concept. California was in the process of changing its procedures in order to centralize claims processing.

The Arizona shared-work program benefited from California's experience and computerized its plan. Furthermore, it was designed to eliminate personal visits of workers on STC to the local UI office or the need for these employees to use the job-search service. Thomas L. Vaughn, the administrator of the Arizona UI Program, indicated that in the second quarter of 1982, administrative costs based on minutes per unit were lower for STC claims than for regular UI claims. Costs may well be further reduced as bugs in this new program are worked out.

Supporters of STC legislation argued that instead of dismissing expansion of STC programs on the basis of increased administrative costs, DOL should be advising states on ways to administer the program to reduce claims costs. These include using the batch claim system by which the employer would file with the UI office for STC for its workers, rather than have the employee appear in person at the UI office. Moreover, work-sharing UI checks could be mailed directly to employees or to the employer for distribution at the work site. More dramatic savings could occur if STC payments were made directly by the employer as in Germany. Finally, the administrative costs in both California and Arizona were overstated due to the failure to consider savings generated because STC workers do not utilize the employment service.

The House Ways and Means Committee was not persuaded by DOL opposition and passed the legislation over Administration objections. Representative Matsui's statement during committee consideration of the bill expressed majority sentiment:

> By all accounts the California and Arizona laws are working well. They are saving jobs that would otherwise be lost and preserving trained and efficient workforces. Other states' reluctance to enact similar legislation seems to stem not from misgivings about the substance of the laws but from a resistance to plowing new ground. It is my hope that passage of my amendment will signal those states that the short-time compensation concept is a sound one and that the United States Department of Labor will help them translate the concept into action."

On July 1, 1982, Senator John C. Danforth (R. MO) introduced S. 2724, an STC bill identical to the Schroeder bill. Although no hearings were held on S. 2724, the Conference Committee, which met on the omnibus Tax Equity and Fiscal Responsibility Act (P.L. 97-248) in August, approved the inclusion in the Act of the STC provisions proposed by the House. P.L. 97-248, signed into law by the President on September 3, 1982, directs the Department of Labor to develop model legislation and to provide technical assistance to states that wish to establish STC programs, to evaluate state STC programs, and to report back to the Congress and the President any recommendations for future policy by October 1, 1985.[22]

NOTES

1. Martin Nemirow, "Work Sharing: Past, Present, and Future," an unpublished paper, United States Department of Labor, 6 May 1976.

2. Alfred and Ruth Gerber Blumrosen, "The Duty to Plan for Fair Employment Revisited: Work Sharing in Hard Times," *Rutgers Law Review* 28 (Summer 1975): 1082-1106.

3. Edith F. Lynton, "Alternatives to Layoffs," paper presented at a conference convened by the New York City Commission on Human Rights, 3-4 April 1975.

4. Edith F. Lynton, Testimony before the U.S. Commission on Civil Rights, Washington, D.C., 1 October 1976.

5. U.S. Commission on Civil Rights, *Last Hired, First Fired: Layoffs and Civil Rights* (Washington, D.C.: United States Commission on Civil Rights, 1977).

6. Ibid.

7. State of California Health and Welfare Agency, *California Shared Work Unemployment Insurance Evaluation* (Sacramento, Calif.: Health and Welfare Agency, 1982).

8. "Leisure Sharing: Is There a Future in Spreading the Work?" Hearings before the Select Committee on Investment Priorities and Objectives, California State Senate, Sacramento, California, 1 November 1977.

9. Memorandum to the secretary on Demonstration Project for Work Sharing Plan, U.S. Department of Labor, 23 November 1977.

10. Martin J. Morand and Donald S. McPherson, "Union Leader Responses to California Work Sharing Unemployment Insurance Program," paper presented at the First National Conference on Work Sharing Unemployment Insurance, San Francisco, 15 May 1981 (*Daily Labor Report*, 28 May 1981, pp. C1, D10).

11. Committee for Economic Development, *Jobs for the Hard-to-Employ* (Washington, D.C.: Committee for Economic Development, 1978), p. 77.

12. National Commission on Unemployment Compensation, *Unemployment Compensation: Final Report* (Washington, D.C.: National Commission on Unemployment Compensation, 1980).

13. *Federal Register*, vol. 45, no. 179, 12 September 1980, pp. 60832–60833.

14. "Part-Time Employment and Flexible Work Hours." Hearings before the Subcommittee on Employee Ethics and Utilization, May 24 and 26, June 29, July 8, and October 4, 1977, Serial No. 95-28, Post Office and Civil Service Committee.

15. "Unemployment Compensation Bills." Hearings before the Subcommittee on Public Assistance and Unemployment Compensation, 26 June 1980, Serial No. 96-113, Committee on Ways and Means.

16. Ibid., p. 161.

17. Work in America Institute, *New Work Schedules in a Changing Society*, a policy study directed by Jerome M. Rosow and Robert Zager (Scarsdale, N.Y.: Work in America Institute, 1981).

18. Richard A. Hobbie, *Potential Effect of Short-Time Compensation on Social Security Benefits* (Washington, D.C.: Congressional Research Service, 1981).

19. Ray Schmitt, *Potential Impact of Short-Time Compensation Programs on Pension Plans* (Washington D.C.: Congressional Research Service, 1982).

20. Ibid., p. 6.

21. "Administration's Fiscal Year 1983 Legislative Proposals for Unemployment Compensation and Public Assistance." Hearings before the Subcommittee on Public Assistance and Unemployment Compensation, March 25 and April 21 and 22, 1982. Serial No. 97-62, Committee on Ways and Means.

22. Texts of the Short-Time Compensation Legislation (P.L. 97-248, Part III, Subtitle G—Unemployment Compensation) and Department of Labor model legislation are in the appendix.

III
SHORT-TIME COMPENSATION: ISSUES AND OPTIONS

10.
Work Sharing, STC, and Affirmative Action

Ruth Gerber Blumrosen

When employers are forced to make a cutback in production, they normally think of laying off some workers. They rarely think about keeping all of their employees at work by reducing the hours of all or many of them. This "automatic reflex" in favor of layoffs instead of reducing hours of work has a long history in the United States and is a practical response in most situations. However, the development of various federal and state equal employment opportunity laws has made the practice of layoff a risky venture. The risks involve the possibility that if the layoff strikes minority, female, or older employees, they will bring a charge of discrimination against the employer, claiming that either no one should have been laid off or that it should have been someone else. Regardless of their outcome, these cases may entangle the employer for years in complex litigation. Under these circumstances, it makes sense for employers to take another look at the possibility of reducing the hours of work of all employees, thereby avoiding the risks of discrimination complaints.

CAN YOU RECOGNIZE ANY PITFALLS?

Test yourself:

1. A private-sector firm, operating under a union contract with a seniority layoff provision, reduces its staff and in the process significantly lowers the percentage of women and/or minorities it employs. As long as it is following a bona fide seniority clause it is perfectly legal to do so.

True   or   False

2. If the union insists on the employer following the contract it is legally secure.

True   or   False

3. If black employees wish to process a grievance under the contract's "no-discrimination-because-of-race" clause, the union may refuse to permit them to do so without legal jeopardy.

True   or   False

4. A public employer has a collective bargaining agreement that contains an affirmative action reference (or statement that in case of conflict the contract will be subordinated to state laws and regulations). The contract also contains a seniority layoff provision. The state has adopted affirmative action guidelines that call for maintenance of minority and female percentages in case of layoff. The employer bypasses women and minorities in its layoff implementation. A white male who would not have been laid off except for the skipping of less senior women and minorities grieves. He will win his grievance.

True   or   False

5. Assuming he loses his grievance, he then seeks to have the union represent him in a reverse discrimination (or age) claim before the Equal Employment Opportunity Commission (EEOC) or its state equivalent. The union refuses. It does so with impunity.

True   or   False

6. The employer raises the "two-bites-at-the-apple" argument in its defense. It wins.

True   or   False

7. The employer (public) argues that it had adopted the affirmative action plan unilaterally, but that the union had become a full partner when it signed the contract. Therefore either the affirmative action plan is *per se* a full defense or the union shares the liability.

True   or   False

8. Abar Corporation, a nonunion company, prides itself on its good employee relations. Every employee on hire is given a copy of the *Employees' Handbook*—a slick publication which points with pride at a record of 75 years without a layoff and explains how the seniority system (based on length of company service), health benefits, and pension systems work. At the end in small print the company reserves the right to modify, amend, or terminate any policy without notice. Subsequently the company announces its first layoff and says it will proceed according to the requirements of the seniority system. Minorities and women who have only recently been hired under an affirmative action plan ask the EEOC and the state antidiscrimination agency to move to halt the layoff on the

grounds that it will have a disproportionately adverse effect on their groups. They will succeed.

True   or   False

9. The result would change if the minorities and women had been hired under the terms of a Conciliation Agreement with EEOC or with the Department of Labor Office of Federal Contract Compliance Programs (OFCCP) or another administrative agency.

True   or   False

10. Even if an employer cannot absolutely rely on a unilateral employer policy or a conciliation agreement, it certainly can effect a layoff in such a way as to protect affirmative action hiring gains made under court order when the court had found liability for discrimination and the employer is trying to avoid being held in contempt of court.

True   or   False

If you answered either true or false to any of the hypothetical problems above, there is a good chance that you are wrong. None can be answered with certainty. The only safe answer to any is "I don't know," "it depends," or "maybe."

For any employer considering a layoff, this massive uncertainty about the state of the law and the possible legal and financial consequences of guessing wrong is not only disquieting—it is indeed the worst state of affairs. Every employer needs as much certainty and predictability as possible in order to plan intelligently. In deciding how to cut back, whether to lay off, and how to design a layoff, most employers in such situations as above could live with any rule so long as it could be relied on. The trouble for employers, unions, civil rights groups, and employees today is that there are no clear-cut rules of the game. Conducting a layoff today is more like blind man's bluff than rational management operations. The present state of the law is such that anyone trying to decide how to make a layoff or reduction in force soon realizes that almost every choice may be buying a law suit.

FOR EMPLOYERS CAUGHT IN THE MIDDLE, IS THERE A DUTY TO WORK SHARE?

The suggestion that there might be a legal duty to work share instead of laying off disproportionate numbers of minorities or women and that employers (and possibly unions) might be vulnerable for "doing a layoff" in the "wrong way" was first made during the 1973–75 recession.

The suggestion was then made that work sharing could be a viable alternative to layoff, particularly if accompanied by payment of prorated unemployment benefits to workers put on shorter workweeks. At that time interest groups were polarizing around the question of how to conduct a layoff. Where employers had only recently begun to hire and promote minorities and women, most labor-management agreements apparently required "last-in, first-out" layoffs that would wipe out the employment gains achieved through affirmative action. Civil rights groups focusing on the loss argued, as do many today, that the affirmative action gains should be preserved and that layoffs should be "proportional" so as to maintain the same percentage of minorities and women after the layoff as were in the work force before. The unions' response was to rally around the preservation of seniority. Seniority is the keystone of the union movement's push for job security and fairness in the workplace. It is the chief mechanism for ending arbitrary employer decisions and allowing the workers to "capitalize their human labor." "Last-in, first-out" preserves the fundamental equities of workers who devote their lives to an employer. When it is applied in a situation where minorities and women have only recently been hired, however, it threatens the gains they have made. Employers were caught in the middle.

With the battle lines thus drawn, Professor Alfred Blumrosen and I[1] suggested that the wrong questions were being asked. Instead of asking *how* to make a layoff, attention should focus on a prior question, i.e., is a layoff necessary? Believing that the congressional objectives in requiring the elimination of employment discrimination in Title VII of the Civil Rights Act of 1964 would be best served by practices that neither subordinated senior whites to junior minorities or women, nor continued to favor white males at the expense of recently recruited minorities and women, we suggested that under some circumstances federal law might *require* some employers to use work sharing instead of laying off workers.

Profile of a Dilemma

In the years since we first made the suggestion, the legal vulnerability of employers (and possibly of unions) for "doing a layoff" in the "wrong way" has become even more complicated and uncertain. This legal uncertainty is a relatively new problem for employers and unions. Since the nineteenth century, employment contracts under American common law have been considered to be employment at will: a doctrine meaning that employees could be discharged for whatever reason the employer chose. This power of management to terminate for "any reason: good reason, bad reason, or no reason" had been largely unchallenged until recently. When employment at will was the rule, an employer could send workers home when work slackened without question. In fact, one of the early arguments against slavery was the greater economy of free labor. The employer had no capital investment in free workers and did not have to support them during slow times. They were simply discharged, possibly with an expectation that they would be rehired when work picked up, but more often without it. During most of this time, management acted as though hiring, training, and discharge were without cost.[2]

Inroads on the Right to Fire at Will

The first inroads on the doctrine of employment at will came through collective bargaining. Unions began to insure some job security by insisting that workers not be discharged except for "just cause" and that proposed temporary interruptions be made in order of inverse seniority. Collective agreements usually also spell out that workers retain a hold on their jobs through recall rights; that is, if production picks up within a stated time, the workers have a right of priority—the right to be recalled according to seniority.[3] The euphemism "layoff" was coined to soften the blow of such firings. Unions generally can enforce their bargain through the contractual grievance procedure ending with arbitration, or may sue for breach of contract. But except for the tentative promise of recalls, laid-off workers are effectively discharged: they lose health and pension benefits while they are out of work, as well as the access to credit that a job provides.

At the same time that collective rights were recognized through protection of the right to organize and bargain collectively, the unemployment compensation insurance system was also instituted. The UI system made employers pay for exercising the right to discharge unless the employee was at fault. Employers were to be encouraged to schedule and allocate resources to avoid unemployment.

A number of statutes also give limited protection against discharge. An employee cannot be discharged because of union activity or race, creed, nationality, religion, sex, or age. Many state laws add other protected categories such as marital and veteran's status. The principle of employment at will has been further eroded in the last several years by a number of court decisions that have found implied contracts not to discharge in employee handbooks and in general obligations of "a good faith and fair dealing," and have held employers liable in tort for discharges that are contrary to public policy or are otherwise "abusive."

These inroads on the employer's absolute prerogative to fire at will are bound to affect notions about the rightness of layoffs. Though most commentators say that lack of work constitutes "just cause" for discharge or layoff, it is not clear that this would be so if there were a viable alternative. Moreover, the criteria for determining if an employer's actions are justified and legal, such as "just cause," "business necessity," "job related," "bona fide," are questions of fact, judgment, and credibility. Most lawyers agree that no one can predict the outcome of a trial dependent on the fact-finder's view of the facts and the credibility of the witnesses.

Compounding the problem are the inherent conflicts among many of the groups protected by this legislation and the fact that the legislation has often been interpreted without taking into account possible conflicts with the objectives and purposes of other legislation.[4]

For many faced with the dilemma, work sharing could provide the answer. Work sharing may be considered a viable alternative even without STC. STC does, however, make work sharing much more acceptable to employees.[5] Work sharing without short-time compensation is as unacceptable to some employees (and their unions) as layoff is to others.[6] STC reduces the risk that employees or unions will challenge work-sharing plans in court.

SOME MORE RISKS AND COUNTERRISKS: CATCH 22 OR DAMNED IF YOU DO AND . . .

An exploration of some of the "Catch 22" conditions under which the layoff question is considered will highlight some of the factors to be alert to and some areas where there are unresolved issues.

Title VII in Hard Times

Under Title VII, the decision to use a last-in, first-out layoff if minorities and women constitute a significant percentage of the newest employees may be illegal if the employer cannot justify the use of the layoff by showing that no viable alternative was available.

The initial application of the equal employment opportunity provisions of the Civil Rights Act of 1964 (Title VII) took place during a period of prosperity. Attention was focused on recruitment, hiring, and promotion practices to ensure that minorities and women would have a fair share of the expanding opportunities. Prior experience had made it clear that minorities and women could not take advantage of expanding employment opportunities without effective enforcement of equal opportunity laws.

In 1971 the Supreme Court said in Griggs v. Duke Power Co. that the *impact* of employment practices, not the employer's *intent*, was where to look for discrimination. *What* the employer was doing that kept minorities or women out was the question, not *why* it was done. Discrimination was defined as conduct that has an adverse impact on minorities or women and is not justified by business necessity, regardless of motive or intent. Under this test, "practices, procedures or tests neutral on their face, and even neutral in terms of intent, cannot be maintained if they operate to 'freeze' the status quo of prior discriminatory employment practices or operate to exclude Negroes . . . [or] operate as 'built-in headwinds' for minority groups."[7]

This broad concept was adopted to assure, as the Supreme Court later said, "that childhood deficiencies in the education and background of minority citizens, resulting from forces beyond their control, not be allowed to work a cumulative and invidious burden on such citizens for the remainder of their lives."[8] The Court recognized "business necessity" as a justification for such otherwise illegal practices, but this concept has been narrowly construed.

The Griggs principle necessarily reaches the employer's planning functions. Duke Power Company's plans to develop a "generally able" work force led it to adopt high school diploma and testing requirements. The Court found that these plans could not justify the adverse impact created by their execution. Thus, wherever an employer practice or procedure is condemned under the Griggs principle and not saved by the business necessity defense, there is a rejection of an employer's plan for carrying out his business and a judgment that the employer could have achieved a business objective without inflicting such harm on minorities and women. Under this principle, an employer who needs to reduce hours of work in order to reduce production or labor costs must carry out the reduction in a manner that does not have an adverse effect on minorities or women, unless justified by business necessity. Thus the employer may be required to assess the probable impact of alternative methods of reducing hours.

Four of the traditional methods of layoff may have the adverse effect proscribed by Title VII.

1. *Last-in, first-out*. This will have an adverse effect if the company (or public employer) has only recently begun to hire minorities or women.

2. *A "selective" layoff, retaining the "most efficient" workers*. This may produce the proscribed effect if it lays off a higher proportion of minorities or women than of other employees, and the employer cannot show that the criteria have been validated under the Uniform Selection Guidelines.[9] Those guidelines require the employer to search for alternatives with less adverse effects as part of the validation process.

3. *Departmental layoff*. If there are still traditional minority or women's jobs and layoff is concentrated so that the primary cuts are made in those jobs, there not only would be an adverse effect, but such a layoff could even be evidence of intentional discrimination.

4. *A two-tier layoff/work-sharing reduction*. If an employer has recently hired minorities or women and institutes a seniority layoff according to

the requirements of a collective bargaining agreement, cutting back so that many of the minorities or women are eliminated and *then* (either immediately or several months later) decides to work share among the remaining (predominately white male) employees, the layoff would probably be illegal. The entire system for reducing working time and labor costs is the "practice" to be considered. This practice had an adverse effect on minorities and women. The difference in treatment between the group that was mostly minorities/women and the white male group probably also constitutes evidence of *intentional* disparate treatment, which probably would not be protected even if the layoff would otherwise have been immune.

If a layoff would produce an adverse effect, an employer may be obligated to plan to achieve the business objectives in some other manner. Obviously a reduction in hours worked by *all* employees would not have a discriminatory impact and is thus a preferable way of meeting the conflicting requirements of business needs, collective bargaining, and Title VII. Work sharing can take many forms, limited only by the creativity of union and management and the exigencies of production. A need for a 20 percent cut could mean going from a five-day week to a four-day week; going to six-hour shifts; laying off employees two days every two weeks; laying off part of the work force for one week (or one month), then calling them back and laying off another segment for a similar period; or utilizing rotating or voluntary time off.

Hard Times and the Older Worker

Under the Age Discrimination in Employment Act, a selective layoff in which the criteria include "potential," "age," "cost," or other criteria that result in the displacement of older employees may, under some circumstances, be illegal.

One of the still unresolved questions under the Age Act[10] is whether the interpretations developed under Title VII in race, national origin, and sex cases will be applied under this act. If they are, an employer practice, policy, or use of criteria or tests that have an adverse effect on older workers is illegal unless the employers can justify their use by showing job relatedness and business necessity. It is argued that this broad "effects test" is required by the Age Act because its language follows that of Title VII. One circuit court has adopted this interpretation.[11]

The counterargument depends on the legislative history of the Age Act, which relied extensively on a report by the Secretary of Labor. The report indicated that even though the language followed the broad language of Title VII, it was not intended that such employer practices as providing for the advancement of younger workers, allowing the employer to keep the lines unclogged, and similar practices that were common in managerial and industrial relations be considered discriminatory and thus illegal, even though they might have an incidental impact on older workers. The Secretary defined only the arbitrary use of age as illegal discrimination, particularly arbitrary age limits on hiring and mandatory retirement. This argument would hold illegal only an employer's intentional use of an arbitrary age between 40 and 70 but would permit employers to lay off the "most expensive" employees, even though they tend to be older workers. An employer who makes a selective layoff that adversely impacts older workers thus runs risks because of the unsettled nature of the law. Employers often lay off both by inverse seniority and by selecting out "the most expensive" or "the least productive" workers. If such criteria are used and many older people are laid off along with inexperienced minorities and women, there is probably multiple liability under Title VII, the Federal Age Act, and state laws. Many states not only prohibit discrimination on the basis of race and sex, but also on the basis of age. Moreover, unlike the federal 40-70 limit, there are no age restrictions in many states, so a layoff that mainly affects young workers (also the ones most likely to be last hired) might be a violation of state age laws. To avoid possible liability under one or more of these laws, an employer might lay off white males with some experience. However, even they may have reverse discrimination or age discrimination claims.

WHEN A COLLECTIVE BARGAINING AGREEMENT CALLS FOR LAST-IN, FIRST-OUT LAYOFFS

If there is a collective bargaining agreement, the contract may require a last-in, first-out layoff. Section 703(h) of Title VII allows an employer to follow the requirements of a "bona fide seniority system."[12] There are at least four areas an employer should investigate before relying on a contract's seniority clause.

A. Is the seniority system "bona fide"? The Court has indicated that a seniority system is bona fide if:

1. It does not have its genesis in racial (sexual, etc.) discrimination.
2. It treats all workers alike and is applied in the same way to both male and female, black and white, minority and nonminority. Therefore, a system that prevented both white and black city truck drivers from transferring to over-the-road jobs was not "intended to discriminate" and was therefore bona fide.
3. The unit in which length of service is relevant was not defined on the basis of race or sex. In the Teamsters case, there were

separate bargaining units for city drivers and over-the-road drivers. Only the length of service in the bargaining unit counted, so drivers could not use seniority as city drivers to bid for over-the-road jobs. The court held the units were not intended to discriminate, even though the effect of the two units was to prevent black drivers from using city driving seniority to bid on over-the-road jobs from which they had been discriminatorily excluded over the years. The bargaining units had been certified by the National Labor Relations Board and had met the board's usual criteria for an appropriate bargaining unit.

4. Whether a seniority system is "bona fide" is a question of fact to be proved by the plaintiff and decided by the trial judge. This means that the trial judge's decision on whether the system is bona fide will usually be the last word on that question. An appeals court will not overturn a finding of fact unless it is clearly wrong. In most cases, the issue will be whether the system had its "genesis in discrimination." This boils down to the judge's assessment of the climate of the times during which the seniority system was introduced and continued, in order to "read the minds" of the participants. The judge will have 20/20 hindsight and is likely, moreover, to apply today's values to events that occurred 20 years ago—not something a careful manager planning a major move would normally want to gamble on.

B. If a seniority system is "bona fide," the question may be whether the layoff is "required." There are circumstances in which a layoff may not be required, including:

1. Over 25 percent of the major collective bargaining agreements provide for some work-sharing possibilities.

2. If the contract language does not explicitly and clearly guarantee specific hours of work to senior employees, it may permit the employer to adopt work sharing. The mere statement that the workweek shall be 40 hours is not necessarily construed to be such a guarantee. Many arbitrators have so held.[13]

3. If the contract is construed as not guaranteeing work to senior employees, would a court be likely to construe Title VII as giving senior employees greater rights at the expense of minorities and women? It is not clear whether 703(h) protects layoffs if management can select employees to be laid off on the basis of ability or criteria other than length of service.

C. Even if the seniority clause is bona fide, and the contract absolutely requires a layoff if management determines that a cutback is necessary, there may be other conflicting commitments. Many employers in both the private and public sectors have entered into conciliation agreements with the Equal Employment Opportunity Commission; government contractors may have agreed with the Department of Labor to take affirmative action in order to retain government contracts, and public employers are also bound by civil service laws, rules, and regulations. When any of these conflict with the seniority clause of a collective bargaining agreement, which will prevail? The Supreme Court has only begun resolving such questions; the answers are still largely unknown.

D. Even if a seniority clause is bona fide and the contract requires layoff, Supreme Court decisions do not necessarily provide protection in cases where contracts were negotiated under state public-sector bargaining laws. In the U.S. Senate debate on Title VII, the national policy of encouraging collective bargaining in the private sector underlies a carefully orchestrated dialogue defining congressional intent with regard to protecting negotiated seniority clauses. The Supreme Court recognized this policy and legislative intent in its decisions regarding bona fide seniority systems. There is, however, no parallel to the case of public-sector bargaining agreements, since equal opportunity laws and executive orders in many states antedated public-sector bargaining laws and could not have been intended to give deference to them. Indeed, many state bargaining laws specifically subordinate the provisions of contracts negotiated under those laws to state laws, regulations, and public policies. The priority that may be judicially assigned to seniority versus affirmative action in public-sector contracts is an open question.

Can a Union Compel a Seniority Layoff?

Even if 703(h) provides a defense to an employer who wants to follow the seniority layoff provisions of a collective bargaining agreement, *must* the employer follow the contract requirements, or may it take unilateral action to avoid the adverse effects of last-in, first-out layoffs on minorities and women? In other words, even if 703(h) provides employers with a shield against claims from minorities and women, does it also give the union a weapon to compel management to follow the contract? The answer to this is complex and unclear. Since the answer is related to several other issues, we will first note some of the other problems that may confront employers, and then explore some of the possible answers.

- Most employees are not covered by collective bargaining agreements. Approximately 70 percent of the work force is not organized. Are *nonunion* employers protected by the bona fide seniority exception if they claim they are following a unilaterally adopted seniority system, which they are not legally bound to follow? The Supreme Court recently said that 703(h) was intended to promote the national policy favoring collective bargaining and indicated that it applies only to collectively bargained seniority systems.[14] The EEOC statement (August 1980) on layoffs indicated that 703(h) protected only employers *required* to go to layoff and to follow the seniority system.

• Whether an employer is required to follow company policy on seniority depends on whether it is legally bound by the unilateral policy. Under the developing law, which is eroding the employment-at-will doctrine, nonunion employers are now caught in a new dilemma. If they advertise the job security and seniority aspects of their employment policies in employee handbooks, or other recruiting materials, they may be held to the seniority provisions and to an implied promise that termination will only be for "just cause." If they do not publicize the company policy, they probably cannot claim protection of 703(h) against suit by minorities and women if a last-in, first-out layoff adversely affects them.

• If employers claim a seniority system protected by 703(h), they may be held liable to long-time employees laid off outside seniority and also sued by minorities and women adversely affected. If they do not claim to be legally bound they may not qualify for 703(h) exception; they will then be liable to minorities and women if a last-in, first-out layoff has an adverse effect or to older people if the layoff is adversely selective.

• Moreover, most statements of company policy reserve to the company the right, unilaterally and without notice, to change, modify, alter, amend, or terminate any policy. Even if such a company policy statement were held to restrict the company's right to fire at will, it probably would not be held to prevent an employer from finding alternatives to layoff, such as work sharing. In fact, lack of work might not even be considered "just cause" if such alternatives were available.

If, in order to avoid the adverse effects of last-in, first-out layoffs on minorities and women or of a selective layoff on older workers, these groups are favored, the decision *not* to follow seniority may have a discriminatory impact on senior white males and result in a suit by them.

Unions Are Also Caught

Unions also face problems in relation to layoffs and seniority. For example: If a union refuses to process grievances from women, minorities, or older workers—or from senior white males—it may be charged with a breach of the duty to fairly represent all members of the bargaining unit. In fact, no matter how a layoff is made, a union may be the target for dissatisfied groups who allege that they were not fairly represented.

WHAT CAN BE DONE?

When Union and Management Agree

If the union and the employer agree to modify or override the seniority clause to protect affirmative action gains, they may prevail in a suit by senior white males charging "reverse" discrimination and/or breach of the duty to fairly represent. The Supreme Court recently provided guidelines to allow unions and management to take collective voluntary action. The Court held that in passing Title VII, Congress intended to encourage employers and labor organizations to voluntarily modify employment practices and remove barriers that adversely affect minorities or women. However, the Court recognized that the interests of white employees also had to be taken into account.[15]

In Steelworkers v. Weber,[16] however, the issue before the Court was whether the union and management could voluntarily consider race in order to open opportunities to minorities from which they previously had been excluded. At least one author has questioned whether opening new opportunities through recruiting, hiring, training, and promotion is really sufficiently similar to the loss of a job to warrant applying the same rules. The guidelines set out by the Court in Weber try to reconcile the conflicting interests. This is done by recognizing the need and rightness of voluntary action—particularly when both union and management agree—but also by severely limiting the scope of the action. The Weber voluntary affirmative action passed muster, even though it disregarded seniority to favor minorities or women because:

1. The purposes of the plan mirrored that of the statute. Both were designed to break down old patterns of racial segregation and hierarchy and open opportunities traditionally closed.
2. The plan does not *necessarily trammel* the interests of white employees. *It does not require the discharge of white employees and their replacement by new black hires.*
3. It did not create an absolute bar to advancement of white employees. Half of those trained in the program were to be white.
4. It was limited. It was a temporary measure, not intended to *maintain* a racial balance, but "simply to eliminate a manifest racial imbalance."

The encroachment allowed is a limited one. No employee was displaced from a present job, put on the street, or deprived of other rights accumulated by length of service. In Weber there was not even defeat of white employee expectations. No white employee at the time of hire could have expected to gain admittance to the craft training program through accumulating seniority. There was no such program. The training program only came into existence because the company and union wanted to take affirmative action to increase the number of black and minority craftspersons. White unskilled workers were the beneficiaries of a program that would not have existed except for affirmative action.

Do Layoffs or Work Sharing Meet the Tests?

How do layoffs favoring minorities or women measure up to these tests? If there is a layoff, someone is going out. The loss of job pension rights, employment prerogatives, as well as related life disruptions, will hit some of the employees. Does this meet the test that no one's rights or expectations should be unduly trammeled if an alternative exists that would have a less severe effect? Work sharing presents none of the forbidding problems associated with layoff. It prevents loss of affirmative action gains and in that sense "is not intended to maintain racial balance but to prevent and eliminate an imbalance"; it avoids *any* adverse effect on *any* group, since all employees share equally in work opportunities. It is limited by the nature of the needs of the business, since all employees are still working, though on reduced schedules. Everyone is available to work full time when production needs increase. Although some expectations based on seniority may be disappointed, particularly for the most senior workers, the impact on the rights of the group as a whole is less when everyone continues to work. Moreover, experience in California, where work sharing with UI has been available for several years, indicates that the impact of work sharing on older workers may not be as severe as a layoff. In a layoff even the most senior workers often must bump down into lower-paying, less desirable jobs for which they must be retrained. With work sharing, all workers stay where they are. The strain of a new learning experience and the uncertainty of who will be next are eliminated.

Moreover, while the expectation that they will continue to work full time may be defeated for the most senior workers, they continue in other respects to enjoy the other values earned by length of service. They can exercise their seniority for purposes of promotion, transfer, shift preference, vacations, parking positions, and other priorities. Most important, the principle of seniority as a method for allocating work opportunities and making choices between employees is left intact. Neither the employer nor the union has to make the often excruciating choice of who must go: everyone stays, though they work shorter hours. No one moves ahead or displaces a senior employee because all employees continue to work at the jobs to which their seniority entitles them. If no further alternatives to layoff are feasible after a time of work sharing, the layoff may still proceed according to the requirements of seniority.

Some people claim this is sharing the misery, but when all share, it is difficult to argue that some are being unduly trammeled. In a world of uncertainty, work sharing seems the safest method of coping with the need to cut back labor costs.

When Management and Union Disagree

Now we are ready to explore whether a union can force an employer to follow the seniority layoff provisions of a contract if management decides to take affirmative action favoring minorities or women. This is another cloudy area, but the EEOC has tried to enable management to prevail over an intransigent union. First, there is an implied requirement to attempt agreement with the union. If such attempts fail or if management knows from past experience that the union will not allow modifications in the contract to accommodate less adverse action, the employer may, according to the EEOC, proceed under the protection of the EEOC Affirmative Action Guidelines,[17] which closely parallel the requirements of Weber.

The guidelines assert that "voluntary affirmative action to improve opportunities for minorities and women must be encouraged and protected in order to carry out the Congressional intent embodied in Title VII. . . . Persons subject to Title VII must be allowed flexibility in modifying employment practices to comport with the purposes of Title VII. . . . Those taking such action should be afforded the protection against Title VII liability which the Commission is authorized to provide under section 713 (b.)(1)."

In other words, the Affirmative Action Guidelines provide a defense to actions by unions or management which comply with the procedure and requirements of the guidelines. Only affirmative action plans or programs adopted in good faith, in conformity with and in reliance upon the guidelines, can receive the full protection. The guidelines:

1. Define circumstances under which voluntary affirmative action is appropriate. Employers, labor organizations, and other persons

   a. May take affirmative action based on an *analysis* that reveals facts constituting real or potential adverse impact, if such impact is likely to result from existing or contemplated practices.

   b. May correct effects of prior discriminatory practices—identified statistically by a comparison between the employer's work force, or a part of it, and an appropriate segment of the labor market.

2. Where, because of historic restrictions, there are circumstances in which the available pool, particularly of qualified minorities and women, for employment or promotional opportunities is artificially limited, persons subject to Title VII may take affirmative action including:
   a. training
   b. recruiting
   c. eliminating unvalidated selection criteria
   d. modification of promotion and layoff procedures, through collective bargaining where a labor organization represents employees, or unilaterally where one does not
3. To establish a plan, three elements are required:
   a. Reasonable self-analysis.
   b. A reasonable basis, including a showing of potential adverse affect, but the analysis need not establish a violation of Title VII.
   c. Reasonable action. Action taken must be reasonable in relation to the problem disclosed. Standards of reasonable action require that the plan should be tailored to solve the problems identified in analysis and to ensure that employment systems operate fairly in the future, *while avoiding unnecessary restrictions on opportunities for the work force as a whole.* A provision conscious of race, sex or national origin should be maintained *only so long as is necessary* to achieve these objectives.
4. Plans will be protected only if they are dated and meet these requirements.

WORK SHARING AVOIDS THE PROBLEMS OF LAYOFFS

With work sharing there are no unnecessary restrictions on opportunities for the work force as a whole—no favoring of *any* group on the basis of race, sex, or national origin or for any other reason. Where work sharing is feasible under these guideline requirements, a race/sex/national origin-conscious proportional layoff would probably be defective and the employer vulnerable to objections from the union or disgruntled disadvantaged employees.

An employer (or union) should be able, in appropriate cases, to rely on the protection of Affirmative Action Guidelines to modify seniority in order to work share. Such protection may not be relied upon if the employer or union goes directly to an affirmative action proportional layoff without first having considered and found insufficient alternatives that would have a less adverse impact on the rest of the work force. These alternatives should include work sharing. Both Michigan and the federal government have adopted affirmative action policies that call for self-analysis and a search for alternatives—probably including work sharing—prior to an affirmative action layoff. Proportional layoff is permitted only if there is no other way to protect affirmative action gains.

Although an employer relying on the guidelines might have a good defense against charges of Title VII violations, it is not clear what the result might be if a union pursued some of the other avenues open to it. The union, for example, might sue for breach of contract if the unilateral management action bypassed the seniority clause; could grieve violation of the contract and go to arbitration; could file a refusal-to-bargain charge with the National Labor Relations Board; or, if the employer is a public body or the federal government, proceed under civil service procedures. Even if the employer ultimately wins, the cost is incalculable, not only in money but in time and attention taken from the primary business of business.

Two more points:

1. Work sharing without supplementary compensation for time lost is unacceptable to many employees. It may be possible for some employers to finance such compensation. Public employers generally provide unemployment insurance (UI) for their employees by reimbursing the Unemployment Compensation Fund on a dollar-for-dollar basis—they are in effect self-insurers. They could probably volunteer to pay an amount equal to a prorated share of UI. It has been suggested that when an employer, by work sharing, avoids having its experience rating raised, it should share the savings with the work-sharing employees. This is an interesting possibility that may help make work sharing more palatable to workers and unions prior to changes in state UI.

2. One repeated word of caution. Often an employer really wants to keep the skilled, experienced part of its labor force and is willing to work share among them. If an employer succumbs to temptation and first lays off by seniority and thus gets rid of minorities and women and then rids itself of unwanted older workers by a selective layoff—and only after these layoffs goes to work sharing for the remainder of the most wanted workers—it is asking for trouble. The difference in treatment of minorities, women, and older workers contrasted with "more desirable," predominately white male workers would most likely be sufficient to make a case of intentional (disparate treatment) discrimination illegal under either Title VII or the Age Act.

Even though we have been talking of work sharing primarily as a reduction from a five-day to a four-day week, employers and state legislators should recognize that this is only one way work may be rescheduled in order to avoid a layoff. Both states and employers should be willing to experiment. Many employers have already introduced innovative and creative ways of allocating reduced amounts of work to keep their present labor force and tie in with the present UI requirements. As the states respond to the federal stimulus and the need to change state UI laws, the focus on a four-day schedule should not obscure the idea that other less orthodox ways of work sharing and tying in to the UI system should also be encouraged. State agencies should refrain from turning down claims of work-sharing employees where, for instance, the employer has decided to rotate layoffs—laying off part of the force for a week or two, then recalling them and laying off another group. Under UI rules, each group would be eligible for compensation, except that in some instances— where the workers had been consulted and had agreed to the new schedule—the state UI agency held they were "voluntarily unemployed" and refused their claim. Such inflexibility should not be allowed to cloud the opportunities work sharing with STC can provide to mitigate the uncertainties as well as the cost of layoffs.

NOTES

1. Alfred Blumrosen and Ruth Gerber Blumrosen, "The Duty to Plan for Fair Employment Revisited: Work Sharing in Hard Times," *Rutgers Law Review* 28 (Summer 1975): 1082–1106.

2. The California Employment Development Department (EDD) evaluation of the California work-sharing program noted the cost of "worker transitions," i.e., when workers switch jobs because of layoffs. If worker A of firm 1 and worker B of firm 2 are laid off and each is ultimately hired by the other firm (worker A ends up working for firm 2 and worker B ends up working for firm 1), both firms would have unnecessarily spent over $3,023 in hiring and training the switched employees, costs that are avoided with work sharing.

3. Most workers who are laid off, however, end up with other employers either because they have no such protection or because their recall rights run out before they can be exercised.

4. Age Discrimination in Employment Act of 1967, 29 U.S.C. 8 261, et seq.; and Alfred Blumrosen, "Interpreting the ADEA: Intent or Impact," in *The Age Discrimination in Employment Act* (Washington, D.C.: Equal Employment Advisory Council, 1982), p. 68.

5. Dissatisfied workers may argue that work sharing to avoid a layoff that would dissipate affirmative action gains is not "reasonable" without STC and that work sharing without supplemental income unduly trammels their rights guaranteed by seniority.

6. The employer tax in the UI system is based on the amount of unemployment charged against an employer's account. This "experience rating" system means that if an employer chooses to work share, it can be done in such a way that employees are ineligible for UI— i.e., in most states employees laid off one day a week are ineligible for UI. In such cases it is argued an employer gets a "windfall," since no unemployment is charged to its account as it would be if there had been an equivalent layoff. Therefore, the whole cost of the recession is put on the workers.

7. The Court adopted the test in Griggs v. Duke Power Co., 401 U.S. 424 (1971). Chief Justice Burger defined discrimination as follows: "Practices, procedures, or tests neutral on their face, and even neutral in terms of intent, cannot be maintained if they operate to 'freeze' the status quo of prior discriminatory employment practices. The Act proscribes *not only overt discrimination, but also practices that are fair in form, but discriminatory in operation.* The touchstone is business necessity. If an employment practice that operates to exclude

Negroes cannot be shown to be related to job performance, the practice is prohibited. . . . Good intent or absence of discriminatory intent does not redeem employment procedures or testing mechanisms that operate as 'built-in head winds' for minority groups and are unrelated to measuring job capability. The Company's lack of discriminatory intent is suggested by special efforts to help the undereducated employees through Company financing of two-thirds the cost of tuition for high school training. But Congress directed the thrust of the Act to the *consequences* of employment practices, not simply the motivation." *Id.* at 430–32 (emphasis added).

On the general concept of discrimination under Title VII, see Alfred Blumrosen, "Strangers in Paradise: Griggs v. Duke Power Co. and the Concept of Employment Discrimination," *Michigan Law Review* 71 (November 1972): 59–110. See also International Brotherhood of Teamsters v. U.S., 431 U.S. 324 (1977).

8. McDonnell Douglas v. Green, 411 U.S. 792, 806 (1973).

9. Guidelines, 29 C.F.R. 1607.5 (1982). In Albemarle Paper Co. v. Moody, 432 U.S. 405 (1975), the Court approved the EEOC Guidelines on Employee Selection Procedures. *Id.* at 431. The Guidelines required that tests must be shown by professionally acceptable methods to be "predictive of or significantly correlated with important elements of work behavior which comprise or are relevant to the job or jobs for which candidates are being evaluated." Few tests are able to meet this strict standard under current validation procedures.

See Michael A. Reiter, "Compensating for Race or National Origin in Employment Testing," *Loyola University of Chicago Law Journal* 8 (Summer 1977): 687–721; James G. Johnson, "Albermarle Paper Company v. Moody: The Aftermath of Griggs and the Death of Employee Testing," *Hastings Law Journal* 27 (July 1976): 1239–1262.

10. Age Discrimination in Employment Act of 1967, 29 U.S.C. 8 621, eq seq.

11. Geller v. Markham, 635 F. 2d. 1027 (2nd Cir. 1980).

12. See Inter. B. of International Brotherhood of Teamsters v. U.S., 431 U.S. 324 (1977). In Teamsters, the Supreme Court held that the Griggs adverse effect test for discrimination did not apply to employer actions taken pursuant to a bona fide seniority system.

13. American Tobacco Co. v. Patterson, 102 S. CT. 1534 (1982).

14. Ibid.

15. United Steelworkers v. Weber, 443, U.S. 193 (1979), U.S., 20 FEP Cases 1 (1979).

16. Ibid.

17. Affirmative Action Appropriate under Title VII of the Civil Rights Act of 1964 as Amended. 29 C.F.R. Sec. 1608 (1979).

11.
Short-Time Compensation: Some Policy Considerations

Martin Nemirow

The simple idea of reducing the workweek rather than the work force has a complicated history, both in the United States and abroad. Therefore, this chapter will look at short-time compensation (STC) in an historical and international perspective. It will also review some important early experiments in the Great Depression—where work sharing without the short-time benefit was used—in an attempt to understand some of the perceptions and misperceptions we now have about the concept. It will also examine some of the possible social and economic benefits that might occur if short-time compensation were ever used widely in the United States. Finally, it will look at some unresolved issues regarding the impact of STC, and the future.*

Work Sharing in the Great Depression

Short-time compensation is a voluntary program to pay temporary partial unemployment benefits to workers on short weeks as an alternative to layoffs. Work sharing in the Depression involved a much different set of economic circumstances. However, it is useful to understand the nation's early experience with work sharing because it has left an emotional legacy of ambivalence that often affects even today's perceptions of short-time compensation.

For example, one feeling expressed is that work sharing was tried by President Hoover and is no better an idea now as short-time compensation than it was then as work sharing. The comparison is instructive. Critics felt that work sharing under Hoover represented an attempt to avoid

---

* The views expressed in this chapter are solely those of the author and do not represent the official views of the U.S. Department of Labor or any government agency.

fiscal or monetary federal intervention as well as to avoid public assistance. Instead, voluntary employer action was encouraged in the form of work sharing, not only to spread the work but to do so without cutting hourly wages. (Hoover felt wage cutting would compound the problem.)

Such work sharing (usually imposed by the employer) subsequently came to be seen by labor as a poor alternative to President Franklin D. Roosevelt's later New Deal measures. Short-time compensation, the current form of work sharing, is a supplement, not a replacement, for macroeconomic policy, transfer payments, and social insurance. There are other differences. President Hoover's work sharing, sometimes used through the early New Deal years, often involved working at half time simply because output was so low. Short-time compensation does not permit employees to work fewer than three days a week and has typically involved four days. Work sharing during the Hoover era was often at poverty-level weekly earnings (there was no minimum wage). Industrial wages are incomparably higher today. Work sharing then was often in unorganized plants (the National Labor Relations Act had not yet been enacted, so unions had minimal power). Today, roughly half of all manufacturing sites, where work sharing has its greatest potential, are organized, and unions would have to agree to STC. And, of course, the Hoover approach did not include partial unemployment insurance, as does STC.[1]

Despite these differences, the fact is that work sharing under President Hoover did save jobs. It seems certain that manufacturing employment might have dropped more than it did in the short term if the workweek had not been sharply reduced from 44.2 to 38.3 hours in 1929–32.[2] This was a 13 percent drop in the workweek, accompanying a 25 percent drop in employment. Since total production decreased by 48 percent, it seems evident that a larger downward adjustment of labor than 25 percent was needed in one form or another. In his memoirs, President Hoover said two million workers had been helped by either work sharing or private relief by employers.[3]

A look at the behavior of the iron and steel industry in that period suggests work sharing's effect in that period. The accompanying graph (fig. 11.1) shows how the employment index had dropped to "only" 50 by mid-1932, although production had dropped to below 20 (1929 = 100). If the workweek index had not dropped sharply from around 50 hours in 1929 to 25 in 1932, unemployment would have been even worse. Note that in the downturn of late 1933 (under President Roosevelt), employment was relatively stable at the same time as hours dropped.

The fact that federal-state unemployment insurance did not exist at that time not only had dire human consequences but also precluded the

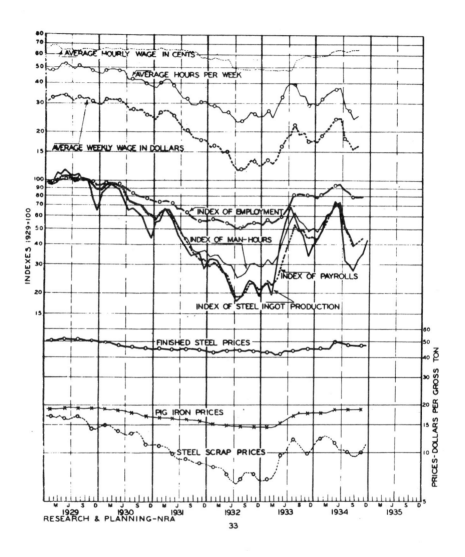

Fig. 11.1. Iron and Steel Industry

Source: National Recovery Administration, *Hours, Wages, and Employment under the Codes* (Washington, D.C.: National Recovery Administration 1935), p. 33.

countercyclical use of UI to offset part of the purchasing power lost by both the fully and partially unemployed.

The second big work-sharing effort came in mid-1933, six months into President Roosevelt's New Deal. The success of voluntary, private work sharing in providing some visible relief had led to demands for more of the same but without weekly pay reductions.

Where President Hoover had tried to prevent loss of some jobs by persuading industry leaders to cut hours, President Roosevelt tried, with some success, to reemploy many of those who had lost jobs by cutting hours still further and establishing minimum wages. His goal was to increase purchasing power while spreading the increased work—all this in a deflationary, not inflationary economy.

The National Industrial Recovery Act, enacted in 1933, was an attempt to increase production, prices, and employment by increasing labor protection and reducing price-cutting competition. (The act also created the Public Works Administration.) It lasted only two years, since it was ruled unconstitutional in 1935. Under the National Recovery Administration (NRA), which administered part of the law, business adopted voluntary codes, including minimum wages and maximum hours. These foreshadowed the Fair Labor Standards Act of 1938.

The NRA actually helped decrease the workweek to 34.6 hours in 1934—it was now 22 percent below pre-1929 levels. This figure reflected a reduction in hours in many low-wage, soft-goods firms from 50–60 hours to 40–48, and even fewer in some higher-wage durable-goods industries. Higher hourly productivity from less fatigued workers, more efficient use of workers, and increased plant utilization accompanied these hours cuts.

Figure 11.2, reprinted from the *Monthly Labor Review* of April 1934, shows how the ratio of jobs to production was increased in part due to NRA workweek reductions, which took effect in mid-1933. The ratio of employment to production was quite low in the pre-NRA upturn of March-June 1933 and much higher in a similar upturn in early 1934.

Specifically, production was at roughly the same level in March 1934 as in May 1933, but a large drop in weekly hours in the interim helped employment (the solid black line) increase even while output dropped. Although it is a subject of debate, some economic historians credit NRA with significant job creation due to work sharing, even while faulting it on other economic and constitutional grounds.[4] The effect can also be seen in figure 11.1. Iron and steel employment was at a similar level in mid-1934 as in early 1930, even though output was much lower in 1934. The difference was the almost 50-hour week in 1930 and the less-than-40 hour week of 1934. The lower fixed costs of that period facilitated such an effect, of course.

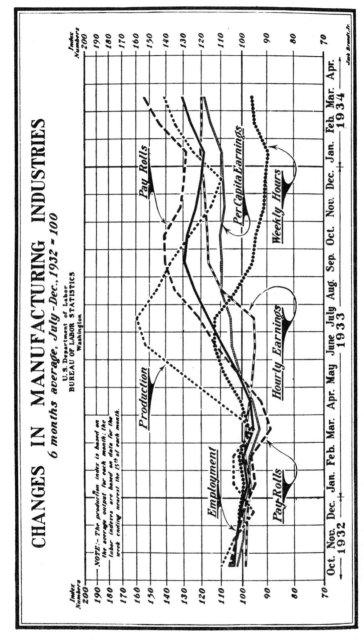

Fig. 11.2. Changes in Manufacturing Industries

Source: Monthly Labor Review, April 1934.

However, the work-sharing effect faded as recovery continued. Due to weak enforcement of NRA, its many exemptions, and finally its demise in 1935, average weekly hours had moved back up to 38.6 by 1937, reflecting hours well above 40 in some firms and much lower in others. Partly as a result, the accelerating increases in output between 1934 and 1937 were accompanied by decelerating increases in employment. While rapid increases in wages may have been one reason behind this increasing gap between output growth and employment growth, longer hours also seem to have contributed.

By 1937, output was back up to its 1929 level, and employment was almost so; however, due to the steady growth of the labor force, some 21 percent of the nonfarm labor force were still unemployed in 1937.

The Fair Labor Standards Act was not passed until 1938; like the NRA, it contained a work-sharing measure in the form of an overtime penalty for weekly hours over 40. The original 1938 ceiling was 44 hours; the 40-hour week was not phased in until 1940. By then, we were rearming ourselves and England, and the effect of work sharing was submerged by the oncoming full employment of World War II.

If work sharing had some beneficial effects in the Depression, why are there so many negative memories of it? One reason work sharing now seems to some a failed experiment of the 1930s is that neither the Hoover nor Roosevelt administrations used modern-day fiscal and monetary measures in a consistent way to deal with the massive unemployment they faced; as a result, work sharing in the 1930s was given a role it could not fulfill, absent these other measures.

To some, President Hoover's work-sharing attempts also symbolized cuts in earnings and the failure of voluntary private-sector-oriented policies to deal with the Great Depression; under the early New Deal, work sharing symbolized to some the unconstitutional approach of the NRA codes. Moreover, both early major experiments in work sharing occurred *before* passage of the Social Security Act of 1935, which brought with it mandatory unemployment insurance and the 1935 Wagner Act (NRA), which gave unions a legal framework for organization (although the NRA also provided the right to organize). With strong unions came strong seniority systems, not only to protect workers against arbitrary dismissal by employers, but to protect them *against unilaterally imposed work sharing*, for it was the practice of many employers not to guarantee a steady amount of work from one week to the next. Workers often showed up at their jobs only to be told there was no work that day. With unemployment insurance came the assurance that low-seniority workers would not starve if they were laid off, and that work sharing, which only "spread the misery," would no longer be needed.

The result of these early experiments *before* all these basic protections for American workers were put into place and integrated with other countercyclical measures (including social insurance and transfer payments) was that work sharing developed a negative image. The current use of work sharing, on a micro rather than a macro scale, is taking place within the framework of these protections and policies. However, the full economic effects of STC, which is a preventive rather than a reemployment measure, in a completely different economy, 50 years later, have yet to be determined.

The Revival of Work Sharing

With the relatively low unemployment rates of the post-World War II era, work sharing was little discussed or used. It was not until the 1974–75 recession, at the time the steepest downturn since the Depression, that work sharing began to be considered again.

In a paper in 1976, I wrote:

> Two major conclusions can be drawn, from a comparison of 1974–75 European with U.S. experience: (1) The portion of unemployment that takes the form of part-time unemployment is much higher than in the U.S. The result would seem to be decreased social costs, increased purchasing power and greater equity, compared to the U.S. (2) The number of U.S. workers who were put on part-time unemployment even in the absence of partial UI benefits is nevertheless not trivial. This suggests that the potential for more work sharing is significant if European-type incentives were instituted.[5]

These conclusions had been reinforced by the New York City Conference on "Alternatives to Layoffs" held in April 1975; representatives of labor, management, and academia reviewed alternatives besides work sharing and found them wanting.[6] Some firms reported mandatory pay cuts, but there was resistance by labor. Cutting health and welfare benefits was ruled out. Voluntary furloughs were found effective by some firms, but they appealed mainly to younger, education-minded workers; older workers nearing retirement; and working mothers. There was also disillusionment about early retirement, due to inflation.

Work sharing was found to be more effective than these other alternatives. However, the case studies presented at the conference pointed up the fact that work sharing without government incentives was usually atypical.

In fact, an underlying crisis for the firm—whereby its very existence was threatened—was a common theme in bringing about work sharing. This was true of Pan Am and the *Washington Star* (the *Star* did go out of business eventually). Union leadership also had to be unusually good in terms of communication with rank and file. (Once unions were convinced the crisis was real, there were often unusual efforts by union leaders to get the rank and file to discuss alternatives to layoffs in day-long meetings and votes.) The firms were often marked by an unusual degree of labor-management cooperation, with management often opening its books. Pan Am went "beyond union contract requirements to develop worker involvement in difficult decisions." Hewlett Packard had an unusually loyal work force due in part to profit sharing.

Nor were the firms especially typical of the average work force. Highly skilled workers were often involved, such as Pan Am flight crews or Newspaper Guild members. It was in the company's interest that young, highly trained people not be lost. There was a team spirit—born of the flight cabin or city room—among the workers. Large numbers of women, many of them the family's second earners, may also have facilitated work sharing in firms such as the New York Telephone Company.

The question was how to create incentives to encourage work sharing in more typical layoff situations as well as in those with the unique chemistry described above.

The fact that these were atypical situations was underlined by a comment of A. H. Raskin, in the *New York Times.*

> For the first time since the great depression four decades ago, unions are finding it necessary to accept pay cuts, work sharing arrangements and other unpalatable accommodations to hard times. [Yet] the dominant view in the AFL-CIO and most of its affiliated unions remains strongly resistant to any retrenchment based on reduction in established contract standards or on retreat from the principle that seniority is the controlling factor in layoffs. . . . The position is that full employment based on tax cuts and other Federal pep pills for the slumbering economy represents the only sound road to recovery—not sacrifices of union gains.[7]

The New York conference found that work sharing in the form of "a shorter work week, or rotating and staggered shifts, or any other method by which average work hours are reduced" emerged as the "alternative to layoffs with the widest potential application to recession-based economic problems and to almost all types of business and industry." It also found, however, that work sharing is "not a panacea. Its use is limited by the necessity of providing a living wage." Thus it found that anything more than a 20 percent reduction in hours would create too much hardship. And when an entire shift must be eliminated, or conversely, only marginal reductions are contemplated—e.g., 20 employees out of 1,000—work sharing would not work.

Appropriateness of STC in the 1980s

The early 1980s have been a period of anti-inflationary restraint, in which our ability to use macroeconomic countercyclical monetary and fiscal policy has been more limited than in the past. Even those who favor micro job-creation tools, such as public jobs programs, usually advocate targeting them to the structurally unemployed—the disadvantaged and long-term unemployed. The disillusion with countercyclical public policies may argue for at least experimenting with policies for the cyclically unemployed that are rooted in the private sector and based to some extent on redistribution of employment rather than solely on countercyclical economic stimuli and public spending.

Moreover, while past efforts to deal with cyclical unemployment have included large public jobs programs, expanded budgets, tax cuts, or new investment, these solutions have not usually had an early impact on recessions or acted as preventatives. To the extent that they have been successful, it has often not been until after, rather than before, layoffs occurred. Job saving has not been a feature of such policies, as it is of STC.

Equity is the major benefit of STC. The economic and social costs of full-time unemployment are distributed more evenly across all workers in a plant (or plant unit) rather than among a small minority of workers.

Some economists have argued that it is the total decline of hours of employment that counts, not its distribution. They see work sharing as a "diversion," a waste of time and resources that could be spent on other countercyclical measures.

However, they may be ignoring the social costs of full-time unemployment. Full-time unemployment increases the costs of public assistance, food stamps, and other transfer programs in a recession. Many studies suggest that full-time unemployment increases the incidence of alcoholism, drug abuse, child abuse, and other social problems, which translate into increased public costs, human costs, and suffering. Distributing the same total hours of unemployment among many people on a four-day workweek might decrease the social costs. It might also help public policy deal in a more rational way with the problem of health insurance for the unemployed, since workers now often lose their health insurance soon after layoff. And, if it were ever adopted on a wide scale, it might also redistribute work and income in a way that bolsters confidence and slows down the decline in consumption in a downturn.

Retraining

STC might also help provide a framework for developing constructive activities, such as education and training, during a downturn. It is unrealistic to think that all workers in a work-sharing program would meaningfully enroll in education or training. However, it would be productive for some. The broad distribution of downtime among the work force would also enable employers to provide part-time training to any workers they feel need it, not just those laid off. (Such training would have to be voluntary on the part of workers, of course.) Public-private mechanisms under the Job Training Partnership Act might conceivably be used for potentially dislocated workers. In general, the meshing of STC with retraining and education is an area deserving further thought.

With hundreds of community colleges and technical schools now operating throughout the nation, it is possible to imagine large numbers of workers who are put on four-day weeks or six-hour days for a two- to six-month period, using that time to attend classes taught on the grounds of the school or inside company walls.

Gunther Schmid, an analyst who has written about employment policy in West Germany, feels that STC is useful as a way of allowing workers to search for work or train for growth industries. He notes:

> The figures . . . make it quite obvious that the real function of short-time work is not restricted to maintaining workplaces and thus keeping or preserving firm-specific human capital. The highest average takeup corresponds generally to those industries which have been hit by structural changes and which reduced their manpower during or after short-time work substantially. Thus, short-time work has had to a considerable extent the function of delaying the reduction of employment which would otherwise have happened like an unavoidable consequence.[8]

It is not clear that STC would be used as extensively by declining industries in the United States as in Germany; the fact is, industrial employment has been declining much longer and more broadly in Germany than in the United States. But it is clear that the United States and other industrial countries are now undergoing a paring down of their work forces in basic industries as they strive to maintain profitability in the face of lower demand and high energy, labor, and capital costs. The idea of a "cycle" in which recovery quickly brings employment back to previous levels in manufacturing is, at this point in western economic history, less applicable than in the past. In this sense, work sharing in the United States may have potential applicability as a transition tool for dislocated workers as well as for cyclical unemployment.

The above are some of the possible national policy benefits in broad brush. There are others, of course, including possible advantages to minorities against layoff and a new degree of cooperation in labor-management relations. (Both of these possible advantages are fully discussed in other chapters of this book.)

It should be stressed, however, that few of these potential benefits are certain. As discussed later on, all need further research.

On a more grass-roots level, California employers who have used STC are clearly enthusiastic about it, as the California evaluation shows; they prefer to keep their work forces intact rather than to lay people off. Layoffs are not only demoralizing but also entail rehiring and retraining costs.

STC also allows business agents of unions to act constructively in a layoff situation, as Martin Morand found in his study of union leader response to California's Work Sharing Unemployment Insurance Program.[9] Generally, layoffs are demoralizing and "no-win" situations for the unions, and business agents do not like helping decide, even indirectly, who will lose jobs. In California, unions generally consulted the rank and file before agreeing to STC and found that the elimination of job insecurity, the spirit of equity, the value of an extra day's leisure (at half or less pay), the savings on work-related expenses—for example, gasoline and meals—all made work sharing popular with workers, before and after it was put into effect.

Need for Evaluation

However, popularity among workers and employers in the short term in one state cannot be the only basis for public policy decisions. A closer examination needs to be made of both possible macroeconomic and microeconomic effects based on a comparison of the economic behavior of workers and employers who use STC versus groups who do not. Only in this way can a worker's short-term gratitude at keeping a job he or she greatly feared losing (by giving up hours of work) be measured against certain economic realities, for example, whether he or she would actually have lost the job. As with any new idea, the task ahead is to explore fully not only its possibilities but any limitations or possible adverse side effects, as well as solutions to them.

Some Unresolved Issues Needing Research

Frank Schiff, vice-president of the Committee on Economic Development and a supporter of short-time compensation since 1975, has summarized some of the key questions that still need to be answered regarding STC. In commenting on the California evaluation, he has noted:

> A question that I believe deserves more intensive exploration relates to the *reasons* for, and the *nature* of, layoffs. To put it differently, more probing is desirable into the question of what the worksharing arrangement was the alternative *to*. Was it an alternative to recession layoffs and thereby in effect kept many people working at least for four days a week who otherwise would not have been able to find any job? Was it an alternative to layoffs that would have resulted in relatively quick placement of workers elsewhere?

Was the arrangement incident to a permanent scale-down in the firm's output? If so, how long should the firm hold on to workers whose basic need is to relocate? Was it an alternative to layoffs at all? [For a number of firms covered, work sharing lasted for a relatively short number of weeks.] Is it possible that for such brief periods of output adjustment, a significant number of the firms involved would have kept *all* their workers on the payroll, had it not been for the availability of the worksharing UI arrangement? Answers to these questions clearly have important implications for the estimates of costs to both private firms and the government. For example, if use of a worksharing UI arrangement in preference to layoffs in effect keeps workers from finding new jobs they might otherwise have obtained quite quickly, the cost of UI to the government would be much higher than otherwise. More information on such points could also help to decide on an important policy issue: namely, should firms be required to present documentation on the need (and reasons for) work-force reductions?[10]

Induced Unemployment and Productivity

One might also add the following questions: Would the added flexibility of a four-day week (allowing retention of a balanced work force) lead to much more use of temporary cutbacks in weekly hours, even to the extent that hours of unemployment might be greater than if layoffs alone were used? Would savings on wages (achieved without losing workers) more than offset the effect of added UI taxes in acting as a disincentive to such added unemployment?

Other possible causes of such induced unemployment may be the fact that employers must claim an equivalent of at least a 10 percent layoff in the affected unit, simply to qualify for STC, even if they were planning less. Moreover, most STC employers probably choose a 20 percent reduction in weekly hours rather than a 10 percent reduction because it saves labor and capital costs for one full day, not because a 20 percent layoff was being planned. Thus, an employer using the entire work force four days a week for three months might only have laid off the number of workers equivalent to the four and one-half day week—or less.

Economists are also concerned about induced unemployment resulting from STC's effect on labor supply. Workers can increase their leisure at a reduced cost, e.g., a 20 percent reduction in work hours with only a 10 percent reduction in income (or less, given resulting lower marginal income tax rates and no work-related cost for the lost day's work).

This improved trade-off between work and leisure is what helps motivate workers to participate, but this reduction in the price of leisure may induce more workers to opt for part-time schedules than are forced to do so by employer decisions to reduce production. To the extent (not clear) that workers and unions could influence employers to adopt part-time work, the question is whether this might increase both UI and production costs.

On balance, one would hope that the job-saving effect of work sharing would outweigh any social or economic costs resulting from most types of induced unemployment, to the extent it exists; that is, we would expect marginal decreases in employment to be less harmful for a group than is total unemployment for part of the group, even if the aggregate hours of unemployment for the whole group are greater. Moreover, the added hours of any induced unemployment may enable an employer to maintain needed productivity and survive at a time of slack production.

However, such questions point up the difficulty of determining the exact job-saving effect. An example of this can be seen in looking at some recent data from Germany.

Table 11.1 shows economic indicators for German manufacturing generally and for the mechanical engineering industry specifically. This industry used STC much more heavily than manufacturing as a whole. Whereas 14.1 percent of wage earners used STC in the mechanical engineering sector (comparable to the nonelectrical machinery sector in the United States), only 10.9 percent of wage earners in other industries

Table 11.1. Percent Changes, 1981–82, in Economic Indicators for Manufacturing and Mechanical Engineering Industry in Germany*

	Manufacturing	Mechanical Engineering
Average Hours	−0.6	−1.6
Employment	−4.5	−3.3
Total Hours	−5.1	−4.9
Output	−2.4	−2.3
Output per Hour	+2.7	+2.6

* German data are based on published and unpublished data of Deutsche Bundesbank, transmitted to BLS on April 14, 1983. The year-to-year changes in average hours consist of changes in holidays and vacations, plus changes in average workweek hours. The great bulk of the change in any short period (e.g., 1 year) consists of changes in the workweek. The decrease in average hours is small, but reflects larger (e.g., 20 percent) decreases in weekly hours multiplied times a fraction of the work force times a fraction of a year. (If, say, 32 percent of the workers were on STC for three months per year, this could be equivalent to an annual average of 8 percent of workers having hours cut by 20 percent—a 1.6 percent decrease in weekly hours). For purposes of this example, we will assume that use of overtime is the same in both sectors, since we are concerned here with the effect on employment of any type of weekly hours changes. Percentages apply to wage earners. Output per hour and average hours are author's estimates. (Bundesbank data are based on official data of German Federal Statistical Office.)

did. We would expect that particular sector to show a much heavier use of short weeks than the typical manufacturing firm and less decline in employment.

Indeed, the table shows that while mechanical engineering reduced total hours about the same percentage (4.9) as manufacturing (5.1), employment was reduced much less (3.3) than in manufacturing (4.5). The loss of employment was more than a third again higher in manufacturing. The decline in average hours was much sharper in mechanical engineering (1.6) than in manufacturing (0.6) and mainly reflects changes in weekly hours.

Note that output per hour (productivity) increased by about the same percent in both cases because total hours were cut back faster than output. In mechanical engineering this productivity increase was due in good part to work sharing augmenting layoffs.

However, is the job-saving effect as dramatic as it seems at first glance? Mechanical engineering has more skilled workers and is less labor intensive. Cost savings from layoffs might be less feasible, making layoffs less likely. Moreover, employers face a greater risk of permanently losing skilled workers. So it is not clear that the mechanical engineering sector would have lost 1.6 percent more jobs in the absence of a 1.6 percent workweek reduction.

In fact, in the 1975 recession, as shown by table 11.2, the U.S. counterpart of this industry (nonelectrical machinery) was unable to cut back hours anywhere near the extent to which it had to cut back output—because it had to minimize layoffs, but did not use extensive short weeks. As a

Table 11.2. Percent Changes, 1974–75,
in Economic Indicators for Mechanical
Engineering Industry in U.S. and Germany*

	U.S.	Germany
Weekly Hours	−1.8	−5.3
Employment	−7.5	−4.5
Aggregate Hours	−9.3	−9.5
Output	−13.8	−5.7
Output/Hour	−4.9	+4.3
Ratio of Change in Total Hours to Change in Output	0.66	1.67

* In the United States, industry is SIC 35; in Germany, industry is Maschinenbau, MAB, DIW 32—Mechanical Engineering. Percentages apply to wage and salary earners. U.S. data are unpublished BLS data that do not meet BLS publication standards.

result, its productivity dropped 4.9 percent. The German industry, by contrast, supplemented layoffs with extensive short weeks and actually increased productivity 4.3 percent. This difference between the U.S. and German industries was due to more hours of unemployment in Germany.

In other words, there seems to have been induced unemployment in the German industry, as compared to the U.S. industry, which accounts for higher German increase in output per hour in a downturn. That this United States-Germany productivity gap is "artificially" widened during a downturn is evident from the fact that the gap for the same industry was nonexistent in growth periods. The U.S. rate was 3.3 for 1970–74 and the German rate was 3.0.[11] This is the rate which measures the real differences in technology and other efficiencies between the same industry in two countries. The 1974–75 gap, therefore, was due in part to the added flexibility in hours cuts afforded by heavy use of short weeks. Did the German industry nevertheless save jobs? Probably, but as noted earlier, not to the extent that weekly hours were cut back. In fact, the decline in total hours of employment in the German industry was 2.5 times greater than in the U.S. industry in relation to output. However, this must be discounted partly by the fact that Germany's decline is a compounding of a long-term decline in employment to a degree that does not exist in U.S. industry. In any case, the German industry's productivity increase was clearly of benefit to the economy.

It is also true that these effects are taking place in a different economy and society and in a UI system that is not experience rated, which probably helps induce hours of unemployment. Nevertheless, even in the United States, there may be a similar effect for some employers: more hours of unemployment; a reduction in layoffs, but not to the extent represented by the cutback in weekly hours; and increased productivity in a downturn.

It seems fairly clear that extensive subsidization is given to employers using STC in Germany, since there is no experience rating, but rather a flat tax on employers and workers (the tax funds other social insurance and training programs, in addition to STC and regular unemployment benefits). The ability of German manufacturers to maintain productivity increases without resorting to extremely heavy layoffs is aided by such subsidization, which takes the form of subsidization by industries not using STC or UI funds as heavily and also subsidization by the federal government in deep and extended recessions.

This may even have implications in terms of international competition. The mechanical engineering industry is an export-oriented one. STC probably helped German manufacturers to compete with U.S. manufacturers in the counterpart U.S. industry (nonelectrical machinery) in 1974–75. (Recent data for the U.S. industry is not available.) As demand declined, the Germans could muster both heavy work sharing and some reduction in force to maintain productivity, allowing them to retain skilled personnel without adding to their UI taxes. U.S. manufacturers not only faced higher UI taxes for whatever layoffs occurred, but also had less flexibility to maintain productivity through work sharing as a supplemental labor adjustment tool. To the extent that some other U.S. industries increased productivity in the 1981–82 downturn, it was due more to layoffs than was the case for counterpart industries in Germany.

One does not have to support the degree of subsidization in the German system to note that strict experience rating of STC in the United States in a recession may maintain that type of international advantage. International trade aside, it may be reasonable to use any infusion of federal funds in a recession (triggered by extended high unemployment rates) not only to assist states in payment of regular UI payments but also to facilitate use of STC. (Of course, the extent of induced unemployment that would occur in the United States needs to be determined before such a policy decision is made.) One possibility would be to use general revenues to compensate senior workers so that they would have the same STC replacement rate as less senior workers in a recession. This would facilitate use of compensation when it is most needed.

Replacement Rates and Equity

Seniority rights are a central feature of our collective bargaining system. While work sharing may promote "horizontal" equity (distributing the burden of unemployment), the seniority system reflects a "vertical" equity principle—that someone who has put in, say, 15 years of work in a workplace is entitled to greater job security than someone who has put in 6 months. These rights have been given legal force through our collective bargaining system and are also recognized in many nonunion firms. They are widely regarded as a protection against arbitrary dismissals by an employer who uses layoffs to get rid of certain employees.

Given the "vertical equity" of seniority, one could argue that there should be at least the same percentage of replacement of higher-seniority wages as lower-seniority wages under any STC program. But in states with low maximums and for workers with high wages, STC asks significantly larger relative financial sacrifices. This would seem unfair, since without STC many senior workers would lose no income at all— i.e., would not be laid off or put on short time. However, bringing them up to the same replacement rate as lower-paid workers when STC is used would add significantly to UI costs. High replacement rates also raise equity questions, since contributions are not based on wages at the high end of the payroll.

John Zalusky of the AFL-CIO, which supports the concept of STC, has given the AFL-CIO's view of the problem:

As it now stands, the unemployment insurance benefit available to workers will be a major disincentive to the use of STC. To get the maximum benefit from STC, the more secure workers will have to *want* to share work with those at the bottom of the employment structure. When one considers the relative sacrifice of workers in a state like Minnesota—selected because it has one of the better ratios of maximum U.I. benefits relative to average wages—it is clear that the more secure workers will be asked to make a significant financial sacrifice for the more easily replaced workers. In January, 1982, the basic maximum weekly benefit was $266 per week. For each day of unemployment elected by these workers, they would receive one fifth of $266 under the typical STC plan or $53.00. The maintenance trades helper (an entry level job) in the Twin Cities, earns about $75 per day or falls short $22 per day. When transportation costs are considered, this entry level worker loses about $3 to $4 per day. By contrast, a maintenance electrician earns $96 per day. Adjusting for transportation savings, they would lose $20 per day (these losses do not consider the one week waiting period). There are no tax advantages since income from unemployment compensation is taxable for all who earn more than $12,000 a year—over $5.76 per hour.

In a state with lower maximum benefits, such as Texas, the disincentive is even higher. The maintenance trades helper in Dallas earns about $64 per day and the usual state STC plan would replace about $29. The maintenance trades helper loses about $20 per day considering work related expenses. On the other hand, the maintenance electrician earning about $91 would have $29 replaced by STC. After work related expenses, the more secure worker will be giving up $39 per work day. This sharing is inequitable since we are asking much more of the secure and more highly paid workers. Thus, the U.I. STC replacement rate should be geared to the individual's wages rather than a state flat rate.[12]

Of course, the U.S. system is quite different from Germany's. German workers contribute to their UI fund, whereas U.S. workers do not. This helps to justify the two-thirds replacement rate that German workers receive, both for regular UI and STC.

The replacement rate issue points up the constraints on the unemployment insurance system to act simultaneously as a work-sharing system. If one had to invent a work-sharing system from scratch in an attempt to replace lost earnings most equitably without increasing costs, a completely different scheme might have been developed. For workers whose jobs are being saved, a very low or no replacement of wages might be given; instead those funds might be redistributed to higher-seniority and higher-paid workers to assure an adequate replacement rate for the earnings they are voluntarily giving up. This would be a work-sharing incentive system rather than a UI system. It would redistribute "work-sharing incentive" benefits upward to the workers who are losing part of their full-time jobs; it would redistribute work downward to workers who would otherwise have none at all.

This is clearly a fantasy, but it points up the difficulties of moving our existing UI system in a direction that requires an adequate work-sharing incentive across the full spectrum of the work force. The purpose of the existing UI system is to insure adequate partial replacement of lost earnings for all workers. Equity within this goal leads to a higher replacement rate for lower-paid workers, compared to more highly paid workers.

Still, there is a wide spectrum of workers whose earnings are in a range that would make them candidates for participation in work-sharing attempts. At this point, there would seem to be room for STC to expand among such workers.

A thornier problem may be that not enough of the lowest paid and least senior workers participate because they are laid off before STC begins.

Martin Morand found that 72.8 percent of union officials said they experienced a reduction in force just before adopting STC. Many of these firms, of course, had not known about STC and had not been able to use it before layoffs occurred. However, others may have known about the program, but decided to lay off their least productive or experienced workers and use STC for the most experienced and skilled workers.

This presents a difficult problem. Even in Germany, where STC is widely used, reduction in force often supplements use of STC, and it is the least productive workers who will most likely lose jobs. This implicates STC as a program that may sometimes discriminate against the workers who need it most; skilled senior workers can most easily find reemployment.

One suggested approach to this problem is to require that only minimal layoffs be allowed during the period prior to STC use, if an employer is to qualify. Even though this may discourage some employers from using STC, any reduction in use would be offset by a broader benefit to the unemployed who need it the most. Others, however, feel this would be unduly restrictive.

Effect on the Total Wage Bill

Logically, cutting the same number of hours from workers across the whole work force decreases an employer's wage bill, when compared to cutting the same number of hours for lesser-seniority and lesser-paid workers. Given the replacement rate problem just discussed, this decrease may be less than we thought at first glance, since relatively fewer higher-paid workers may join. However, 67 percent of employers in California cited savings in labor costs as a reason for using STC, and such savings are probably part of those overall labor-cost savings.[13]

The social benefits and costs of such a lower wage bill deserve further examination. On the one hand, unit labor costs are being decreased at a time when an employer's profits are being squeezed. This helps the firm to survive. Even if profits are not sharply declining, lower labor costs, in theory, allow more investment, leading to future growth and jobs. Also, such wage bill savings may be the key to offsetting the higher fixed costs of work sharing. On the other hand, if STC ever became widespread, too large a savings in the wage bill could decrease purchasing power as well as tax revenues and dampen the spending that is needed for economic recovery (thus inducing unemployment in yet another way)—especially because the replacement rate is not as generous as in Germany. (Germany's high replacement rates may cause other problems, however, such as a disincentive to work.)

Given the relatively limited use of STC, however, the benefits to the firm and its workers will probably outweigh any such macroeconomic effects on consumption or taxes.

Triggers and STC

Some economists might also argue that in a period of economic growth STC would lead not only to induced unemployment, but that it would be inflationary to assist a given employer to hold on to his workers by allowing STC. In a growth period, jobs are available that could be filled if the worker were let go. Labor mobility might be cut by STC at just the time when it is beneficial to the economy in terms of keeping wages at no higher than their equilibrium level. Wage demands, which are strong anyway, would be even stronger, and wage restraint would be less tempered by the knowledge that layoffs could follow unreasonable increases in labor costs.

Wage increases in Europe in the 1970s (higher than in the United States), during periods of growth but high unemployment, may indeed have been exacerbated by use of STC with very high replacement rates and other job security measures. Complete job security and immobile labor markets in a growing economy are a formula for inflationary wage increases that accelerate substitution of capital for high-cost labor, leading to slow, long-term employment growth. This may be in part what has happened in Europe, where industrial employment has declined over the long run, while in the United States it has expanded.

In a recession, slack labor markets allay such concerns. If further experience and research show there is substance to these or other fears about STC use in normal times, steps could conceivably be taken to make STC easier to use when the economy seems likely to head down. National leading indicators rather than the unemployment rate might be used as a trigger, since preventive action should occur *before* widespread layoffs and consequent unemployment occur.

Making STC harder for an employer to use in a nonrecessionary period would, of course, be an automatic way of screening out seasonal industries or other employers with questionable economic rationales, although it would also screen out dislocated workers from declining industries unless special exceptions were made.

The issue of a trigger does not arise in Germany because the German UI system requires economic evidence of the need to use STC and screens out seasonal use of STC. There is also greater worker input into decision making, through more widespread unions and works councils. Experience rating may serve much the same purpose in the United States, but that is not completely clear.

Windfall Effect

How many workers have their weeks shortened even without STC? How many more would be added with STC? If the latter is significantly greater than the former, then STC is of value in increasing work sharing. Otherwise, STC becomes mainly a partial UI benefit—equitable in itself, perhaps, but of less importance with regard to unemployment policy.

If we estimate the increased use of short weeks in Germany during the last two recessions and compare it with the increased use of short weeks in the United States, we get a fairly constant ratio: In Germany short weeks were used 2.5 times as much as in the United States. In the absence of more refined research at this point, we might use this ratio as the potential expansion of the U.S. STC program. Before discussing this further, let us note the derivation of this ratio.

Between 1980 and 1982, there was an increase by 469,000 of German workers on STC, while unemployment increased by 944,000, a ratio of 0.50 STC recipients to unemployed. In the United States, in the same period, there was an increase of 460,000 workers who had been working full time but were put by employers on part time (less than 35 hours per week). The ratio of this group to the increase in U.S. unemployed (2.4 million) was 0.19. Dividing the German ratio by the U.S. ratio gives us 2.6. About the same ratio results when these calculations are done for the 1973–75 downturn in the two countries.[14]

The 2.5 to 1 ratio might be regarded as a very rough estimate of potential expansion. To get a more refined estimate would involve controlling for the different industry mix in Germany as well as the many other variables. In addition, there are other statistical refinements that would need to be made in comparing the two countries. But this ratio, for the present, may serve as a framework for discussion. Our sense that we are in the right order of magnitude is reinforced by the fact that the Bureau of Labor Statistics' international comparative data indicate that in both 1973–75 and 1979–82, about two-fifths of Germany's decrease in total hours of employment was comprised of decreases in average weekly hours, compared to about one-fifth for the United States. Since overtime is a larger proportion of the U.S. weekly hours cuts, we would expect the use of below-normal workweeks to be more than twice as great in Germany.[15]

Interestingly enough, Martin Morand found that 47.3 percent of union leaders in firms using STC said their firms used work sharing even before STC was available. Of course, they may have used it for shorter periods of unemployment, but here again, we have roughly twice as many firms using work sharing as did so without STC.

Thus, on the one hand, existing use of short weeks is not insignificant in the United States. (We cannot "prove" that they save jobs, but that is also often true of STC.) For those concerned about windfall, there were 460,000 workers who would be candidates for STC even though their workweeks were cut without STC. Of course, with a voluntary, experience-rated program, the adverse effect of such a "windfall" is not clear.

On the other hand, if we are willing to use the German-U.S. ratio as a rough outside limit for the United States (in a recession), any such "windfall" would be only part of a significantly larger work-sharing effect, if Germany is any clue.

At this stage, it is not possible to estimate whether the United States could actually achieve a "take-up" rate comparable to Germany's or whether it would be far less. Certainly, there are many factors that favor work sharing's proliferation in Germany more than here, such as high benefits, no experience rating, and more pressure not to lay off workers. If one combines such German variables with the fact that some U.S. employers who now use short weeks without STC may actually switch to a layoff mode because their UI tax savings will be lost, one might speculate that the use of short weeks might, at most, double in a recession rather than increase one and one-half times. In 1980–82, there would thus have been another 460,000 workers on short weeks.

Assuming no induced unemployment, the 460,000 more workers on short weeks in 1980–82 would have saved the jobs of 92,000 of those workers. That is a significant number that would result from implementing STC in every state. On the other hand, it is only 4 percent of the increase in the 2.4 million unemployed in the period. Of course, the increased purchasing power from the total number of persons collecting STC would also have helped create jobs.

What about nonrecessionary times? In the United States in 1979, there were 1.25 million part-time workers who ordinarily would have worked full time, while unemployment was 6.0 million—a ratio of 0.21. But in Germany, short weeks that are compensated for by STC fall way off in a growth year. Thus, only 88,000 workers collected STC in 1979, even though 876,000 were unemployed, a ratio of only 0.10. Data are not available to indicate whether other German workers were put on part time without STC, as in the United States. However, even if they were, it is evident that the German government feels there is much less economic justification for such payments when the economy is growing.

Other Issues

Other policy-related research that needs to be done includes finding the true cost to firms of workers who find jobs after layoff. Is it the case that most firms choose temporary layoff because they know most workers will be available for recall? Will this be even more true in coming years now that so many firms have cut back blue-collar work forces and laid-off workers have far fewer comparable jobs to which they can turn?

What is the true social cost of the layoffs that are averted by STC? Are such relatively short, temporary layoffs actually responsible for much of the psychological and financial stress that sends child abuse figures up and increases the need for welfare and social services? Or are such costs mainly the result of permanent layoffs and structural unemployment that offer a worker little hope? If so, is there a realistic strategy for encouraging use of STC in these situations to foster job search and training?

These and other questions are yet to be answered. The California evaluation, while useful, did not have a control group of employers, nor were hard data available on many of the issues discussed here.

Conclusion

In its present embryonic state, the future of STC is hopeful but not certain. The goal of sharing the burden of unemployment is easier to state than it is to implement, and there are still a variety of questions regarding its effects and optional use.

The fact that STC has not caught on in the United States as it has in Europe can be explained, in part, by the difference in the evolution of U.S. and European labor-market institutions. Workers in Europe have been more collectivist and egalitarian than in the United States.[16] Sharing the burden of unemployment has been more acceptable to them and strict seniority is of less concern.

Employer subsidization of certain benefits for the unemployed that are lacking in the United States (for example, health insurance) means that the added fixed costs of work sharing for employers in Germany are less. The costs of layoffs, in terms of severance pay and prenotification by employers, are also greater, as are the legal obstacles to layoff. European labor, management, and government are also more centralized than in the United States, and labor unions have more direct political power, making STC easier to implement. And, of course, the STC benefit is more generous.

Another key difference is the much greater control German unions and works councils have over possible employer abuses in order to help assure that there is economic justification for short hours.

Despite all these differences, the United States has periodically experimented with work-sharing approaches and this is a period of such experimentation. This trial period may help develop a useful supplemental tool to deal with unemployment.

One key to STC's wider use may be whether U.S. managers develop a more detailed understanding of the long-term as well as short-term costs and benefits of layoffs as compared to work sharing. Frank Schiff has suggested the need to develop a sophisticated cost and benefit accounting system to provide managers with data to make such decisions. In the atmosphere of a business downturn, such data rarely is developed in a way that comprehensively deals with each firm's hiring, training, benefit, and productivity costs. Such a system might help broaden decision making beyond what is sometimes a fixation on today's fixed costs of labor.

As business awareness of the need for long-term planning increases, such tools will hopefully emerge. But beyond such technical tools may lie a more fundamental issue, productivity. The general consensus among experts in the field of worker productivity is that efforts to increase labor productivity through worker participation, quality circles, and other devices depend heavily on a worker's sense that increasing productivity will not lead to his or her loss of a job. STC would seem to offer a potentially valuable tool for employers, to help maximize job security in return for better long-term productivity performance in the United States.

NOTES

1. Irving Bernstein, *The Lean Years: A History of the American Workers, 1920–33* (Boston: Houghton Mifflin, 1960), pp. 306–307 and 476–484, provides a detailed account of Herbert

Hoover's President's Emergency Committee on Employment, which urged voluntary work-sharing efforts.

2. All historical data in this section are from the Bicentennial Edition of Historic Statistics of the United States, U.S. Department of Commerce, Bureau of the Census, Parts I and II, September 1975. Production data (FRB) p. 667, labor force and unemployment data, pp. 126–7; manufacturing employment data, p. 137, earnings and hours data, p. 170, productivity data (NBER), p. 162; CPI data, p. 210. Data in these volumes are from CPS and BLS unless otherwise noted.

3. Herbert Hoover, The Memoirs of Herbert Hoover: The Great Depression, 1929–41 (New York: Macmillan, 1952), p. 45.

4. "The President's Re-Employment Agreement gave jobs to about 2,462,000 persons between June and October 1933 through reducing weekly hours of work. . . . Industrial activity in this period declined, hence, the increase in employment was the result of shorter hours. . . . However, NRA codes, after they substantially superseded the President's Re-Employment Agreement, added very little to the number of jobs between October 1933 and the first five months of 1935, in spite of a gain in manufacturing production of 14 percent . . . due to tolerances, exceptions and exemptions. In 64 percent of the codes, covering 61 percent of employees in codified industries, provisions permitted a workweek of 48 hours or longer for many of these workers. The abuse in the application of loosely drawn provisions reduced the reemployment." From Broadus Mitchell, "Depression Decade," in volume IX, The Economic History of the U.S. (New York: Rinehart, 1947), pp. 283–284.

5. Martin Nemirow, "Worksharing: Past, Present and Future," unpublished paper, 6 May 1976.

6. Edith Lynton, "Alternatives to Layoffs," a paper prepared for a conference convened by the New York City Commission on Human Rights, 3-4 April 1975.

7. New York Times, 9 March 1975, sec. 3, p. 1.

8. Gunther Schmid, "Selective Employment Policy in West Germany: Some Evidence of Its Development and Impact," Special Report No. 27, European Labor Market Policies, National Commission for Manpower Policy, Sept. 1978, p. 202.

9. Martin J. Morand and Donald S. McPherson, "Union Leader Responses to California's Work Sharing Unemployment Insurance Program," paper presented at the First National Conference on Work Sharing Unemployment Insurance, San Francisco, 15 May 1981.

10. Comments on the evaluation of the California work-sharing UI program by Frank Schiff at the First National Conference on Work Sharing Unemployment Insurance, San Francisco, 14 May 1981.

11. German data, including Table 11.2, published by Deutsches Institut für Wirtschaft Forschung, Berlin, May 1980; U.S. data from Bureau of Labor Statistics.

12. Remarks by John Zalusky to the Interstate Conference of Employment Security Agencies, Inc.

13. State of California Health and Welfare Agency, California Shared Work Unemployment Insurance Evaluation (Sacramento, Calif.: State of California Health and Welfare Agency, 1982), p. ES-3.

14. 1982 German data are from "OECD Economic Survey of Germany," June 1983, Table I, p. 76. U.S. data on part-time workers are from the Monthly Labor Review, table 3, "Selected Employment Indicators, Seasonally Adjusted, December 1982, p. 59, and February 1983, p. 61.

15. Annual average changes in average hours and employment for Germany and the United States are given in the U.S. Department of Labor news release of 26 May 1983, "International Comparisons of Manufacturing Productivity and Labor Cost Trends—Preliminary Measures for 1982."

16. Everett Kassalow notes that the trade union movement in the United States is not a class movement as in Europe and that the "class-oriented union movement in any country

[including Europe] dislikes the role of helping to determine who will be laid off [by means of seniority clauses]. Seniority is given some consideration when layoffs are made, but factors like family responsibilities, degrees of skill, and the like may count for as much or more." In some ways the absence of seniority provisions "reflects the greater industrial power and job control in American unions [compared with Europe]." In Europe, the author implies, such control is not felt to be needed, since widespread acceptance of trade unions in Europe gives them less need to build in defenses against possible arbitrary layoffs or promotions through detailed seniority clauses. From Everett M. Kassalow, *Trade Unions and Industrial Relations: An International Perspective* (New York: Random House, 1969), p. 142. These observations were written, of course, before the high European unemployment rates of recent years.

12.
STC and Labor-Management Cooperation

Ramelle MaCoy
Martin J. Morand

"In America when sales drop the first thing American management thinks of is laying off employees. In Japan it is the last resort." So says Paul Hagusa, the president of a Japanese corporation, Sharp Manufacturing Company of America, operating in Memphis, Tennessee.[1] "There must be a change in the way that management in the U.S. treats its employees. Otherwise it will be difficult for employees to respond," Hagusa continues, "[because] . . . the most important path to quality is that employees feel they are part of one family, playing an important role in the company."

As American management ponders the advice of Mr. Hagusa and other Japanese industrialists, it is easy to forget that the "quality circle" concept that is believed the key to Japanese productivity and quality is a notion originally introduced into post-World War II Japan by an American industrial engineer. American managers tend to treat it as an exotic foreign notion whose successful transplantation into our environment can only be achieved with great difficulty.

Alarmed by the success of foreign competition, management has sought to duplicate foreign management methods. The names given the programs initiated by individual companies and the consultants they retain are varied. The meanings attached to the names are even more varied, with usage dependent on the company that adopts them and the context in which they are applied. Titles include quality circles (QC), quality of working life (QWL), labor-management participation teams (LMPT), and labor-management committees (LMC). Close cousins are relationships by objectives (RBO) and employee stock ownership plans (ESOP), though these efforts to encourage workplace cooperation are at least partly government sponsored or encouraged by tax legislation.

In the brief span of this chapter it would be impossible to dispel all of the confusion stemming from the profusion of names and initials, but in an effort to avoid compounding the confusion we shall use one term, labor-management cooperation (LMC—used herein only with reference to labor-management *cooperation*), to generically denote the full range of activities that promote improved relations between management and a unionized work force. Of course, much of what we report will have application to the nonunion firm as well.

Labor as a purely variable cost—one that can be reduced through layoff at will—is a peculiarly American notion. On the other hand, "partnership" and "family," concepts which relate closely to labor-management cooperation, also have deep roots in the American experience. Management's new focus on LMC arises out of an effort to create those harmonious relations that are thought to have given foreign companies a competitive edge in both productivity and quality. Although some American workers and their union leaders have responded favorably, others have frequently reacted with some skepticism, feeling that if they were truly "partners" or "family" they would scarcely need to be informed of the fact by management.

More important, the use of these terms in the absence of job security is likely to continually raise questions in workers' minds that increase the initial skepticism. How do you fire a partner? Who ever heard of laying off a family member? Differences between Japanese (or European) and American concepts of job security have tended to be ignored as American firms have attempted to build on foreign blueprints for cooperation without considering the foundation on which industrial cooperation abroad is based.

If trust and sharing are to be more than a passing fad, and if American labor-relations practitioners are to learn anything from other systems, they will have to recognize the centrality of job attachment to these systems. In order for American workers and their unions to undertake the risks inherent in LMC, it is essential for them to feel secure in their jobs.

The Japanese system of "lifetime employment" will be hard to transplant. It rests on social bases probably unacceptable in our own society. Even in Japan, only a minority of workers enjoy the status of lifetime employment that includes relative class and employment immobility. Women are almost totally excluded from the employment guarantee; retirement is compulsory and takes place at an earlier age and with less adequate pensions than in the United States; and employees generally show a degree of commitment to employers that Americans would find unacceptable. Yet these are essential parts of the Japanese style of management.

More acceptable approaches may well be found in Europe where workers also have a far higher degree of job attachment and security than in the United States. The relative immobility of labor in Europe may be partly cultural and partly the product of more homogeneous societies. But to a large extent it is a reflection of public policies that restrict plant closures and layoffs, encourage early retirement, and subsidize employment stability with funds otherwise designated for employment benefits.

STC, a slight variation on existing unemployment insurance law, not only encourages a work-sharing alternative to layoffs but may surprisingly and unexpectedly contribute as much to the actual practice of LMC as more elaborate programs such as quality circles and quality of working life. While the primary objective of STC is to maintain an entire work force's job attachment, in practice the implementation of a work-sharing plan, in virtually every case we have examined, has been found to have had a beneficial impact on employer-employee and union-management relationships.

If not solutions, STC at least provides alternatives to, or avoidances of, many of the problems typically associated with efforts to move from competition to cooperation without cooptation. First and most important, STC immediately eliminates one of the major obstacles to employee-union-employer cooperation—job insecurity.

The identification of a connection between security and cooperation is not a foreign ideology nor is it of recent vintage. The idea that safety is a precondition for social behavior is not an invention of Maslow's.[2] It was emphasized by Slichter in 1941, Golden and Ruttenberg in 1942, and the importance of security for workers, employers, and unions continues to be the focus of recent works such as those by Bluestone in 1981 and Work in America Institute in 1982.[3] German "co-determination" and Japanese "quality circles" are only new names given to old and very American ideas.

Layoffs Create Labor-Relations Problems

The impact of layoffs on unions, management, and the relationship between the two can be illustrated by an example.[4] The labor relations director at a large California manufacturing plant says that during a major layoff, typically involving 500 to 1,000 workers, there will inevitably be hundreds of grievances, even though he tries to comply scrupulously with contractual provisions relating to layoff. Most of the grievances are ultimately won by the company, but only after a great deal of time and energy have been expended in documenting the accuracy of records and the appropriateness of judgments. Some of the grievances that are lost by the company, the labor relations director claims, arise because he has been relatively successful in negotiating a layoff provision that is *not* just a wall-to-wall, last-hired, first-fired system. Like many seniority clauses, it is qualified, with the right to bump an employee into a lower job classification that requires prior experience in the job and the ability to perform it at a satisfactory level. Shop-floor records of such skills and experience are generally imperfect.

When the company loses a grievance, it is usually after extensive and time-consuming meetings and frequently as the result of costly arbitration. Reinstatement not only means a back-pay cost but a disgruntled supervisor who has, in effect, been told he or she was wrong and an angry "other" worker—the one who has now been displaced.

From the perspective of the union business agent at the plant, the problem is, if anything, worse. He felt that management generally made a good-faith effort to abide by the contract and was convinced that many of the management judgment calls being challenged by displaced workers were probably correct. But his role and function, legally and politically, was as advocate for the disadvantaged, even when the disadvantaged did not suffer illegal or discriminatory treatment. Because the union is a political organization, the union representative cannot afford to be perceived as an apologist for management. He must be an advocate for the worker. And in recent years, a series of court decisions rapping the knuckles and the pocketbooks of unions that failed to give workers what they conceived of as "fair representation" have made such a posture not only politically wise but legally necessary. More and more unions are processing questionable grievances to arbitration simply as insurance against the costly duty of fair representation litigation.

Almost always, as in this California plant, a major layoff, particularly where recall is not imminent, gives rise to a flood of grievances filed by those selected for layoff. They have nothing to lose, and the union business agent in practice has little option but to battle for the grievants and engage in the adversarial meetings, arguments, and hearings that are apt to poison any cooperative spirit that may have existed and shape antagonistic postures that may endure for years.

For the union, even a victory is often Pyrrhic. Workers who are reinstated feel that they received no more than was due and indeed may even give the credit for the reinstatement to the arbitrator. Workers who have been displaced are often more disgruntled and bitter than had they been picked for layoff in the first place and sometimes successfully sue the union, claiming that it failed to adequately investigate and represent their right to retain employment.

Sharing Without Unemployment Compensation

Of course, there is and always has been an alternative to layoff by simply sharing the work by reducing the hours of each worker without any unemployment compensation supplement. But such a "solution" to avoiding layoffs creates far more problems than the one it was designed to solve.

Workers who have their paychecks slashed by the reduction of the workweek by a day or two object to the loss of pay. But they resent even more bitterly being cheated out of what they perceive as a benefit due them by right. The often heard comment: "They give you just enough work to screw you out of your unemployment benefits" is a reflection of a sense that a legislated fringe benefit that is perceived to be part of the total employment package is being withheld.

The union will frequently grieve and carry to arbitration charges that the employer, by unilaterally reducing the workweek, violated a contract containing a seniority layoff provision and/or an hours of "regular work" clause. Neither union nor management can be confident of the outcome of such cases, and unpredictability, uncertainty, and instability are not conducive to a harmonious relationship. One recent study of hundreds of arbitration cases involving work sharing versus layoff finds:

> There is great variety in arbitration decisions in this area. . . . Arbitrators disagree as to whether worksharing is to be considered the same as laying off or not . . . management retains the right to schedule work unless restricted by the collective bargaining agreement . . . there is no consensus as to the specific limits on management's right to change the work schedule by spreading the available work among all employees instead of laying off by seniority."[5]

Layoffs and Discrimination

The grievances, arbitration, litigation, and threats of litigation that either conventional layoff/bumping procedures or unilateral workweek reductions give rise to obviously create conflicts between union and management representatives that are counterproductive to efforts to establish constructive cooperation. A related and even more abrasive effect of the last-hired, first-fired layoff procedures arises because such systems frequently result in a disparate impact on women, minorities, and other protected classes. If members of such classes have been recently hired, perhaps as the result of affirmative action programs, a seniority-based layoff may lead to charges of discrimination. And even if such charges are ultimately not sustained, the charge itself may poison the labor-relations atmosphere.

When Equal Employment Opportunity Commission (EEOC) investigations are undertaken in response to such charges, there is frequently a fearful and defensive posture on the part of management, the union, or both, depending on the thrust of the complaint. As in the grievance and arbitration situation described above, winning most cases and losing only a few is of little solace. The implications of the conflict over "rightful place" are much greater in cases involving protected classes than where the conflict is over "only" individual rights. When racism or sexism rear their heads in the workplace, they undermine efforts toward LMC. A work force and union membership divided against itself cannot effectively cooperate with management.

Layoff and the Worker

The problems that layoff can cause to the union and management when they are attempting to create a more facilitative *modus vivendi* go beyond these institutional problems of grievances, arbitration, and litigation. The economic, social, psychological, and physiological impact of unemployment on the individual worker poisons his or her (particularly "her," according to one recent study)[6] relationship with the union as a member, with the employer as a worker, and possibly the union-management relationship itself.

The Trouble with Layoff: A Management View

The president of a small manufacturing company summed it up thus:

> Suppose I've taken a young guy with a few years service and laid him off for several months. I've told him, "Fend for yourself. How you pay your mortgage and feed your family is your problem. If you or your wife or kids get sick, forget about medical insurance. It's over. If you need help, go to the government for unemployment benefits or food stamps. If you start drinking, quarreling with your wife, and beating up on your kids, call some community agency. Whatever your problem is, don't call me!"
>
> Several months later business picks up. I call him back. If he returns (some of the best workers will have gone elsewhere), he returns with the drinking problem, the self-hatred and the loss of self-esteem which unemployment creates—and he is all too ready to transfer those angers toward me. It is a statistical certainty that he is more likely to develop health problems, mental and physical, while unemployed than while working, and those illnesses will ultimately manifest themselves in premium increases in my experience-rated health insurance program.
>
> Immediately his return is disruptive to production. Just at the moment I am anxious to make a quick response in filling new orders, I am forced to reverse the bumping procedure and move everyone around like pieces on a chessboard. This is more than a technical disruption. The members of my work force do not operate as individuals. They work in teams, and every

time that I change the composition of a work group it is not only the new member who must learn the routine. The entire new group requires time and practice to function together. It is not just a physical synchronization; it is social and psychological as well. It takes months, possibly years, to optimize a work team's pace, skills, and internal relationships.

One might assume that after the traumatic experience of unemployment this guy would come back grateful to have a job and thus be a more loyal employee. Forget it! He has learned one lesson and he may never forget it: this is not *his* company. He's not even a junior partner. He is a variable labor cost easily dispensed with when times get tough. We kept paying our creditors and maintaining our machines. We seldom lay off managerial or white-collar workers. But when he needed us most, we weren't there. He has no sense of security or stability with this organization. He may seem to forgive. He never forgets.

Except for wages themselves, layoff is one of our most expensive labor costs. The guy I'm talking about may not even consciously get mad, but he will unconsciously get even. The anger may express itself through written grievances that he would formerly have shrugged off or a vote against contract ratification. More likely it will manifest itself in poor morale, poor workmanship, absenteeism and even accidents. We simply are not a profitable enough enterprise to afford to lay off workers we expect to need again. That's why I prefer a shared-work-time reduction, with our unemployment comp taxes and the money we would have spent on layoff and turnover used to maintain our workforce.

The Trouble with Layoff: A Union View

For the union rep or officer, the problems created by the displaced worker are different but no less difficult. The problem starts, one local union officer explained, even before the layoff itself.

When workers see the orders slowing down and the inventory building up, they smell a layoff. The work pace slackens, machines malfunction, rejects rise, and supervisors get upset. There are threats of discipline, arguments as to whether work errors were intentional or accidental, and generally a lousy atmosphere throughout the plant. Those members whom we had assigned to work with management on productivity improvements, waste reduction, and quality control are suddenly accused by their fellow workers of being the cause of the layoff problem. "You've efficiencied me right out of a job," they say.

When people start going out the door, things get worse. Aside from the problem of priority—who goes and who stays and on what job—there are new complaints both from those working and those unemployed. Most of our members depend on incentive earnings for a good part of their paycheck. The bumping not only immediately reduces the hourly job rate for some people, but it disrupts the entire plant so that incentive earnings drop while workers are adjusting to new jobs. Those still working are not sure for how long and it makes everybody edgy.

The member who has been paying dues for years is suddenly faced with the biggest job-related problem he's ever had. He comes to you for help and you tell him to go to the unemployment office for help. It sure doesn't increase his respect for you or for the union.

For some of the guys here work wasn't just a paycheck. It was a major part of their lives. So they came around the plant at shift change or lunchtime just to be with the gang—but they wind up bitching. They hang around the union hall and want to know when I'm going to get them back to work. Maybe they've heard that some supervisor has done some work on their machine and they want to know what I'm going to do about it. If I say anything that suggests the company may not be completely in the wrong, they bristle. "Whose side are you on?" they want to know.

For some reason many workers see their health insurance as something they got for being a union member, something that their dues paid for. When they lose these benefits, they lose some of their belief in the union. We try to keep in touch with those on layoff, to offer help and advice about government and community services. But they are a bitter bunch and a few of them would be willing to dump the union or undermine the contract if they thought it would get them back to work. Some of our guys got sent to a nonunion plant during layoff, and they actually bad-mouthed the union during an organizing campaign—as if it were the union that laid them off.

When they start being recalled it's even worse. They all think they ought to be next. I'm constantly in and out of the personnel office checking personnel records, seniority dates, and time cards. It's a no-win situation.

And it doesn't settle down and blow over that quick when we do get back working. Getting along with management is a risky road for a union officer in the best of times. Layoff is the worst of times.

Short-Time Compensation and Labor-Management Cooperation

STC is not competitive with LMC. In fact we consider it to be the necessary precondition to any real and long-term LMC. In addition to serving as a foundation for other forms of LMC, STC is itself a form of LMC which has many intrinsic advantages.

The adoption of an STC plan may provide, quite unintentionally, an ideal opportunity and environment for the introduction of an LMC program. Indeed the cooperative labor-management discussions and decisions necessary to the implementation and administration of an STC plan appear to have led naturally, in some cases, to a *de facto* unplanned and unnamed "quality" program.

STC is introduced to alleviate an immediate, short-term, specific problem. There isn't enough work to go around, and that fact is usually obvious to everyone. STC raises no great expectations and makes no large promises. It does not ask that management and unions make profound changes and adopt an alternate way of dealing with each other. Rather, it structures a situation that by its nature requires and nurtures cooperative decision making for the allocation of the same unemployment benefit dollars in a different way.

By definition, STC is a limited, short-term program. Not only does legislation usually limit its utilization to a 26-week period, but studies of workers, employers, and unions found all much more eager to try it for a 13- or 26-week period than to make a commitment for a longer period.[7]

It is noteworthy that STC provides a congressionally authorized framework for entering into a relationship that, according to evidence, is likely to generate positive relations among workers, between workers and their employers, between workers and their unions, and between unions and employers.

STC, overtly and with legal authorization, deals with several questions that LMC committees are probably proscribed from discussing. In a unionized setting, STC requires by law the union's cosignature to the firm's application to the state agency. There can be no question of end-running the union.

STC necessitates dealing with schedule changes. The legal question as to whether this is a mandatory subject for bargaining or merely for discussion is ignored. The parties both discuss and bargain about what to do about the program. They do so securely because each holds veto power—there can be no STC plan unless both agree.

STC necessarily goes beyond traditional bargaining for wages, hours, and working conditions and does so in a completely natural way. The decision to adopt the alternative work-time reduction requires, logically and casually, discussions of the cause of the problem and how long it is likely to last. Sales projections, competition, revenue, markets, and investments may all be seen as relevant to the discussions, as are potential cures for the firm's distress.

Worker Involvement and the STC Decision

A review of the California experience has demonstrated a very high degree of shop-floor participation prior to the initial decision to try STC. Because it was a variation from the normal requirements of the contract, the union representative would not and the personnel director could not institute the change unilaterally. A tremendous amount of education about this alternative to layoff took place in the workplaces investigated. Usually it received overwhelming support. In one instance where it ran into significant opposition, the labor and management leaders did not move ahead with the program.[8]

The trusting, sharing, and open communications that are identified as essential components of QWL and similar programs are equally essential within the union if it is to take the risks and try for the benefits of cooperation. The union leader seeking to change an historically antagonistic relationship with management will have to convince a skeptical or hostile membership. Where LMC may be too big a pill to swallow, STC may prove to be just the right size to try.

In one situation, a firm and a union entered upon an STC program following a long history of acrimonious relations, including picket-line confrontations and litigation going all the way to the U.S. Supreme Court. The initial worker reaction was one of amazement. Commented one worker: "He [the employer] cares enough to try to keep us working!" This shock treatment literally shook up the relationship. The employer was heard to compliment the workers on the better quality of their workmanship during the reduced work schedule, and they responded by working even better.[9]

The study of the California experience led to the conclusion "that one of the major reasons for the generally positive response to this program is that the [union] members were asked and not ordered to participate."[10] Union business agents reported that a strategy had to be developed to deal with psychological and political realities in order to secure worker approval.

One union representative reported that he was always careful to select the natural leaders of the most likely objectors and meet with them to explain the STC concept. Another union representative explained that anytime a modification of the contract with respect to seniority was considered, he called a meeting of that portion of the local union most likely to be affected by work sharing and nearly always secured a unanimous vote. One business agent reported the tactic of presenting information at a joint meeting on company premises, after which the company representatives were excused so that the vote could take place privately.[11]

STC Leads to LMC on the Job

STC generates understanding between workers and their employers, workers and their unions, and unions and employers. As one representative of a large international union put it, "management got good marks from the workers just because the programs showed that they cared." A business agent for another union reported "the program fostered a deep awareness among workers of the business of the company."[12]

From the perspective of some union representatives, STC is a barometer of labor relations; the program only works where relationships are reasonably good, and it improves them when it works. One union representative noted that if workers believe they are being treated unfairly by the employer, "workers will cut off their noses to spite their faces. Suppose you are a senior forklift driver working four days and a less-senior forklift driver in another department where work is still available for five days gets a fifth day, which nets him $8 or $10 more than you make after taking into account taxes, travel, and the partial unemployment compensation. Would you really want to work a day for a dollar an hour? No, not unless something else was bugging you. But if there is trust, petty grievances don't get filed."[13]

That this cooperation comes not by preaching alone, but out of the day-to-day experiences with problem solving was evidenced by one company's experience with work sharing on a weekly rotational basis.[14] Their system of alternating two weeks work with one week off was designed to take advantage of the unemployment compensation in the absence of a state STC law. It led to certain measurable improvements: less waste, higher productivity, and fewer grievances. For the first time in many years a contract was negotiated without a strike and ratified weeks before the expiration date of the prior contract. More significant: when work sharing itself was insufficient to cope with the firm's business problems, the sharing experience had created a cooperative environment in which the union and its members were willing to try further experiments at mutual problem solving for mutual survival, and they tried an ESOP.

Labor-Management Cooperation in the Political Arena

There are additional positive LMC impacts that grow out of the political process of passing and implementing an STC law. Policy statements in favor of STC have been issued by the Committee for Economic Development,[15] the AFL-CIO,[16] and Work in America Institute, a work research organization supported by both business and labor.[17] Testimony in favor of such legislation has come from leading corporations and unions. State legislation on STC has been sponsored by Democrats and Republicans with equal frequency. While there are many instances of labor and management agreement as the result of compromise, only one other form of social legislation (for the establishment of health maintenance organizations) has received such positive and enthusiastic support from employee and employer representatives. In the early years of the California program, "the enemy of my enemy is my friend" concept held sway as both the union and the management of a major plant marched to Sacramento to protest what they felt were unfair rulings and poor administration. They jointly and successfully prevailed upon the Employment Development Department to change the rulings and upon the state legislature to modify the legislation to overcome the problems they had both identified.

STC and Flexitime

LMC often attempts to deal with flexibility in scheduling. There is a traditional union resistance to flexitime because it tends to undermine hard-won daily overtime provisions. One of the products of STC is a heightened awareness of the value of leisure time both personally and economically. Because the definitions of the "work units" that may engage in STC are flexible and a uniform reduction in hours for the entire enterprise is not required, employers and unions have learned to custom-tailor schedules to take into account individual worker preferences and employer needs. An outgrowth of an STC experience may become a continuing application of voluntarism in adapting both total hours worked and working schedules to individual needs and preferences—a development that would contribute greatly to humanization of the workplace and to LMC generally.

STC and Security—for the Union and Its Leaders

The security of the union as an institution has been identified as an essential goal if LMC is to work. STC is reported to "assist in maintaining union membership and strength." For one business agent, this was the key advantage to STC, since, in his view, "a reduction in numbers means a reduction in strength."[18] Just as STC contributes to job security and stability, it also contributes to union security and therefore to the stability of the labor-management relationship.

One sophisticated official felt that any program that improves relations between workers and the employer, with the union getting credit, has to positively benefit the union. He reported on the improvement in interpersonal relations among the workers: "Workers like each other and like the boss. When workers hate the boss and their jobs," he added, "they also hate the union."[19]

In conclusion, it is appropriate to acknowledge that there is no evidence that STC is the wave of the future, any more than LMC may be. Nor do we advance STC as a substitute for LMC. STC is primarily a labor-market adjustment mechanism and only incidentally and accidentally a vehicle that may contribute greatly to the growth of LMC. In fact, this unplanned side effect may, in time, overshadow the purpose for which short-time compensation was created.

NOTES

1. *The Wall Street Journal*, 29 April 1983, p. 1.
2. Abraham H. Maslow, *Motivation and Personality* (New York: Harper, 1954).
3. a. Sumner H. Slichter, *Union Policies and Industrial Management* (Washington, D.C.: The Brookings Institution, 1941).

   b. Clinton S. Golden and Harold J. Ruttenberg, *The Dynamics of Industrial Democracy* (New York and London: Harper & Brothers, 1942).

   c. Irving Bluestone, "Quality-of-Work-Life Goals Fulfill Union Objectives," *World of Work Report*, December 1981, p. 91

   d. Work in America Institute, *Productivity through Work Innovations*, a policy study directed by Jerome M. Rosow and Robert Zager (Scarsdale, N.Y.: Work in America Institute, 1982), pp. 146–151.

4. Martin J. Morand and Donald S. McPherson, "Union Leader Responses to California's Work Sharing Unemployment Insurance Program," paper presented at the First National Conference on Work Sharing Unemployment Insurance, San Francisco, 15 May 1981 (*Daily Labor Report*, 28 May 1981, pp. C1, D10).

5. Michael J. Haberberger, "The Arbitration of Worksharing Disputes: Development and Implications," *Labor Arbitration Information System: Perspectives* 10 (May 1983).

6. Kay A. Snyder and Thomas C. Nowak, "Impact of the Robertshaw Shutdown on the Psychological Well-Being of Terminated Men and Women," paper presented at a meeting of the Pennsylvania Sociological Society, Millersville, Pa., 13 November 1982.

7. Zeena Weber, Joseph T. Sloane, and Robert D. St. Louis, *Shared-Work Unemployment Compensation: Arizona Survey Results* (Phoenix, Ariz.: Arizona Department of Employment Security Unemployment Insurance Administration, 1981).

8. Morand and McPherson, "Union Leader Responses."

9. Ibid.

10. Ibid.

11. Ibid.

12. Ibid.

13. Ibid.

14. Linda Roberts, "Worksharing: A Rotational Furlough Plan Saves Jobs at McCreary Tire," *World of Work Report*, February 1982, p. 9.

15. Committee for Economic Development, *Jobs for the Hard-to-Employ: New Directions for a Public-Private Partnership* (New York: Committee for Economic Development, 1978), p. 77.

16. Statement by the AFL-CIO Executive Council, 5 August 1981.

17. Work in America Institute, *New Work Schedules for a Changing Society*, directed by Jerome M. Rosow and Robert Zager (Scarsdale, N.Y.: Work in America Institute, 1981), pp. 88–98.

18. Morand and McPherson, "Union Leader Responses."

19. Ibid.

Afterword

Ramelle MaCoy
Martin J. Morand

It is too early to predict with confidence the future of STC in the United States. Early successes of the program in California, Oregon, and Arizona as well as abroad—particularly in Canada—would seem to guarantee its extension, at least on an experimental basis, into other American states. The costs of the program are so minimal and the possible benefits so great that it would seem difficult for its opponents or detractors to deny it at least a trial.

Nevertheless, ten months after passage of federal legislation mandating that the Department of Labor "assist states" in developing such laws, STC had been adopted by only two states, Florida and Washington. The Florida initiative came from Motorola and the Associated Industries of Florida; Washington was inspired by its neighbor, Oregon. Neither was responding to any federal suggestions or assistance.

It took the Department of Labor ten months to produce the model legislative language which the new law had instructed it to develop, and during that period some states were still copying sections from the 1978 California law that California itself was reconsidering and revising in the light of experience. (The original 20-week limitation on STC benefits was increased by California to 26 weeks and a punitive surtax on negative-balance employers was eased sufficiently to permit, in practice, the participation of such employers.)

Federal legislation also mandated that "the Secretary [of Labor] shall conduct a study or studies of State short-time compensation programs consulting with employee and employer representatives in developing criteria and guidelines. . . . " It is regrettable that one year after passage of the legislation, neither consultations nor studies had taken place.

The lack of enthusiasm for STC displayed by the Department of Labor to date is difficult to explain and stands in sharp contrast to the attitudes and actions of the Canadian government. In 1982 Canada initiated a modest STC program with a budget of $10 million, but by year's end the positive response and experience with the program led to an increase in the 1982 budget to $200 million, with $250 million allocated for 1983.

Canada's initial 26-week program was extended to 38 weeks during the first year, and in 1983 a 50-week program was made available to firms anticipating a permanent reduction in their work forces. The 50-week program was coupled with a subsidized worker-employer retraining and placement program for permanently displaced workers.

While the similarities between the Canadian and U.S. economies and societies prompt the conclusion that a comparable commitment to STC by the United States would lead to comparable results, there are at least two striking differences that give STC an edge in Canada. The Canadian national health insurance system makes STC more attractive to a Canadian employer since it reduces the cost of health insurance benefits for extra workers retained. STC is also more attractive to Canadian workers since the Canadian unemployment system pays a higher maximum benefit than any U.S. state. In addition, the Canadian STC program requires no waiting week and does not offset STC benefits against total UI entitlements in the event of ultimate layoff.

One of the explanations for greater Canadian government interest in, and acceptance of, work sharing may lie in that government's awareness of the high social and health costs of unemployment. A preliminary evaluation of the Canadian STC program (*A Preliminary Evaluation of the Work Sharing Program, Employment and Immigration Canada*) found "reduced expenditures on health and other social programs," and when these savings are taken into account the cost/benefit ratio of STC increases significantly to 1:5.7 (see chapter 8, page 116).

It is possible that adoption of pending congressional proposals for health insurance for the unemployed together with similar initiatives at state levels might lead to a comparable U.S. awareness of the health costs of unemployment. Should either the federal or state governments adopt programs that require the payment of health benefits or insurance for the unemployed, there would be a more direct and immediate incentive to subsidize at least part of the STC employer's additional costs for health insurance as in Germany.

Our citations of the success of STC in Canada should not suggest that we feel that comparable successes are certain in the United States or that STC is a cure-all for our economic ills. We do, however, see it as much more than an emergency measure. Its greatest ultimate benefit may prove to be the positive effect of security on productivity.

Experts in the field of worker productivity, Martin Nemirow reports in Chapter 11, generally agree on this proposition: "Efforts to increase labor productivity through worker participation, quality circles, and other devices depend heavily on the worker's sense that increasing productivity will not lead to his or her loss of a job. STC would seem to offer a potentially valuable tool for employers, to help maximize job security in return for productivity performance."

Skeptics may find these and other potential benefits of STC too glowing to be entirely credible, but the fact that the costs of the program are modest is beyond reasonable dispute. Essentially, STC amounts to little more than the legislative authorization of a rather minor change in UI procedures to permit the same benefit dollar to be distributed in a slightly different way. The financial burden of the program is borne predominantly by participating workers, all of whom—except for those who would otherwise have been laid off—experience a reduction in earnings. The available evidence suggests, however, that American workers themselves look upon the loss in earnings as a reasonable price to pay for the job security and increased leisure that the program provides.

Whatever cost there is to employers—and the only significant costs are those attributable to the expense of maintaining fringes for a larger number of employees—appears to be judged by virtually all participating employers to be amply repaid through the elimination of turnover costs, maintenance of productive capacity, and increased worker morale and productivity.

The costs to government are more difficult to measure. There might be, but certainly do not have to be, slightly increased administrative costs. There might also be some reduction in revenues from graduated income taxes because of the lower levels of earnings of participating workers.

It should be borne in mind, however, that the greatest contribution that the unemployment insurance system has made to society, generally, has been the maintenance of purchasing power to prevent layoffs from generating additional layoffs. There is, however, an important practical economic distinction between purchasing power and effective consumer demand. The fear and insecurity that permeated our society during the severe recession of the early 1980s at times resulted in a curious phenomenon: an increase in personal income which occurred simultaneously with reduced consumer purchases, as workers hoarded wages against feared unemployment.

While regular UI benefits are clearly as effective as STC in maintaining purchasing *power*, we strongly suspect that STC is more likely to maintain effective consumer *demand*. If this is true, then the increased tax revenues from such purchases and the related higher level of economic activity may more than counterbalance any reduction in income tax revenues.

The greatest benefit of STC may ultimately turn out to be subtle but dramatic changes in the attitudes of employers, employees, and unions toward each other. The unstated assumption implicit in the adoption of an STC plan is that the employer-employee relationship is a valued and continuing association. The implication is that the partners in that relationship are pledged to work cooperatively to overcome temporary difficulties.

We think it will be unfortunate if attempts to evaluate STC focus too narrowly on immediate costs and benefits and fail to explore positive attitudinal changes as well as other indirect and peripheral benefits that seem to be generated. As the attempt is made to calculate the net worth of STC, we suggest that the following questions be considered:

(1) Does STC, by maintaining employment and engendering confidence, sustain consumer purchasing more effectively than regular UI?

(2) To what extent does STC improve and encourage labor-management cooperation?

(3) To what extent does STC reduce the social costs of layoff, including child neglect, alcoholism, drug addiction, divorce, illness, and delinquency?

(4) What is the optimally efficient workweek for various businesses, industries, and job classifications?

(5) What is the real cost of layoff-related turnover to various industries and businesses?

(6) How effective is STC in avoiding conflicts between minority rights and seniority rights?

(7) What is the extent of administrative savings attributable to the fact that STC claimants have no need to utilize the job service?

(8) How accurate are assumptions of a correlation between seniority and wage rates, and what is the impact on STC benefit levels?

(9) Will STC create pressure for, and economic justification of, a permanently reduced workweek?

The answers to some of these questions must remain subjective. Widespread experimentation with STC, however, is certain, whatever its other values, to very rapidly provide a mass of data upon which to base objective answers. The questions themselves are suggested by anecdotal evidence from employers and employees in California and elsewhere who, having experienced work sharing with STC, think they have had a glimpse of the future and are convinced that it works.

Appendix

I. THE FEDERAL SHORT-TIME COMPENSATION LAW: A SECTION OF THE TAX EQUITY AND FISCAL RESPONSIBILITY ACT OF 1982
(September 1982)

Sec. 194. (a) It is the purpose of this section to assist States which provide partial unemployment benefits to individuals whose workweeks are reduced pursuant to an employer plan under which such reductions are made in lieu of temporary layoffs.

(b)(1) The Secretary of Labor (hereinafter in this section referred to as the "Secretary") shall develop model legislative language which may be used by States in developing and enacting short-time compensation programs, and shall provide technical assistance to States to assist in developing, enacting, and implementing such short-time compensation program.

(2) The Secretary shall conduct a study or studies for purposes of evaluating the operation, costs, effect on the State insured rate of unemployment, and other effects of State short-time compensation programs developed pursuant to this section.

(3) This section shall be a three-year experimental provision, and the provisions of this section regarding guidelines shall terminate 3 years following the date of the enactment of this Act.

(4) States are encouraged to experiment in carrying out the purpose and intent of this section. However, to assure minimum uniformity, States are encouraged to consider requiring the provisions contained in subsections (c) and (d).

(c) For purposes of this section, the term "short-time compensation program" means a program under which—

(1) individuals whose workweeks have been reduced pursuant to a qualified employer plan by at least 10 per centum will be eligible for unemployment compensation;

(2) the amount of unemployment compensation payable to any such individual shall be a pro rata portion of the unemployment compensation which would be payable to the individual if the individual were totally unemployed;

200

(3) eligible employees may be eligible for short-time compensation or regular unemployment compensation, as needed; except that no employee shall be eligible for more than the maximum entitlement during any benefit year to which he or she would have been entitled for total unemployment, and no employer shall be eligible for short-time compensation for more than twenty-six weeks in any twelve-month period; and

(4) eligible employees will not be expected to meet the availability for work or work search test requirements while collecting short-time compensation benefits, but shall be available for their normal workweek.

(d) For purposes of subsection (c), the term "qualified employer plan" means a plan of an employer or of an employers' association which association is party to a collective bargaining agreement (hereinafter referred to as "employers' association") under which there is a reduction in the number of hours worked by employees rather than temporary layoffs if—

(1) the employer's or employers' association's short-time compensation plan is approved by the State agency;

(2) the employer or employers' association certifies to the State agency that the aggregate reduction in work hours pursuant to such plan is in lieu of temporary layoffs which would have affected at least 10 per centum of the employees in the unit or units to which the plan would apply and which would have resulted in an equivalent reduction of work hours;

(3) during the previous four months the work force in the affected unit or units has not been reduced by temporary layoffs of more than 10 per centum;

(4) the employer continues to provide health benefits, and retirement benefits under defined benefit pension plans (as defined in section 3(35) of the Employee Retirement Income Security Act of 1974, to employees whose workweek is reduced under such plan as though their workweek had not been reduced; and

(5) in the case of employees represented by an exclusive bargaining representative, that representative has consented to the plan.
The State agency shall review at least annually any qualified employer plan put into effect to assure that it continues to meet the requirements of this subsection and of any applicable State law.

(e) Short-time compensation shall be charged in a manner consistent with the State law.

(f) For purposes of this section, the term "State" includes the District of Columbia, the Commonwealth of Puerto Rico, and the Virgin Islands.

(g)(1) The Secretary shall conduct a study or studies of State short-time compensation programs consulting with employee and employer representatives in developing criteria and guidelines to measure the following factors:

(A) the impact of the program upon the unemployment trust fund, and a comparison with the estimated impact on the fund of layoffs which would have occurred but for the existence of the program;

(B) the extent to which the program has protected and preserved the jobs of workers, with special emphasis on newly hired employees, minorities, and women;

(C) the extent to which layoffs occur in the unit subsequent to initiation of the program and the impact of the program upon the entitlement to unemployment compensation of the employees;

(D) where feasible, the effect of varying methods of administration;

(E) the effect of short-time compensation on employers' State unemployment tax rates, including both users and nonusers of short-time compensation, on a State-by-State basis;

(F) the effect of various State laws and practices under those laws on the retirement and health benefits of employees who are on short-time compensation programs;

(G) a comparison of costs and benefits to employees, employers, and communities from use of short-time compensation and layoffs;

(H) the cost of administration of the short-time compensation program; and

(I) such other factors as may be appropriate.

(2) Not later than October 1, 1985, the Secretary shall submit to the Congress and to the President a final report on the implementation of this section. Such report shall contain an evaluation of short-time compensation programs and shall contain such recommendations as the Secretary deems advisable, including recommendations as to necessary changes in the statistical practices of the Department of Labor.

II. MODEL LEGISLATIVE LANGUAGE TO IMPLEMENT STATE SHORT-TIME COMPENSATION PROGRAMS. PREPARED BY THE U.S. DEPARTMENT OF LABOR

A. *Definitions*

1. "Affected Unit" means a specified plant, department, shift, or other definable unit consisting of not less than __ employees to which an approved short-time compensation plan applies.

2. "Fringe Benefits" include, but are not limited to, such advantages as health insurance (hospital, medical, and dental services, etc.), retirement benefits under defined benefit pension plans (as defined

in Section 3(35) of the Employee Retirement Income Security Act of 1974), paid vacation and holidays, sick leave, etc., which are incidents of employment in addition to the cash remuneration earned.

3. "Short-Time Compensation" or "STC" means the unemployment benefits payable to employees in an affected unit under an approved short-time compensation plan as distinguished from the unemployment benefits otherwise payable under the conventional unemployment compensation provisions of a State law.

4. "Short-Time Compensation Plan" means a plan of an employer (or of an employers' association which association is a party to a collective bargaining agreement) under which there is a reduction in the number of hours worked by all employees of an affected unit rather than temporary layoffs of some such employees. The term "temporary layoffs" for this purpose means the separation of workers in the affected unit for an indefinite period expected to last for more than two months but not more than one year.

5. "Usual Weekly Hours of Work" means the normal hours of work for full-time and permanent part-time employees in the affected unit when that unit is operating on its normally full-time basis, not to exceed forty hours and not including overtime.

6. "Unemployment Compensation" means the unemployment benefits payable under this Act other than short-time compensation and includes any amounts payable pursuant to an agreement under any Federal law providing for compensation, assistance, or allowances with respect to unemployment.

7. "Employers' Association" means an association which is a party to a collective bargaining agreement under which the parties may negotiate a short-time compensation plan.

B. *Criteria for Approval of a Short-Time Compensation Plan*

An employer or employers' association wishing to participate in an STC program shall submit a signed written short-time compensation plan to the Director for approval. The Director shall approve an STC plan only if the following criteria are met.

1. The plan applies to and identifies specified affected units.

2. The employees in the affected unit or units are identified by name, social security number and by any other information required by the Director.

3. The usual weekly hours of work for employees in the affected unit or units are reduced by not less than 10 percent and not more than ___ percent.

4. Health benefits and retirement benefits under defined benefit pension plans (as defined in Section 3(35) of the Employee Retirement

Income Security Act of 1974), will continue to be provided to employees in affected units as though their work weeks had not been reduced.

5. The plan certifies that the aggregate reduction in work hours is in lieu of temporary layoffs which would have affected at least 10 percent of the employees in the affected unit or units to which the plan applies and which would have resulted in an equivalent reduction in work hours.

6. During the previous four months the work force in the affected unit has not been reduced by temporary layoffs of more than 10 percent of the workers.

7. The plan applies to at least 10 percent of the employees in the affected unit, and when applicable applies to all employees of the affected unit equally.

8. In the case of employees represented by an exclusive bargaining representative, the plan is approved in writing by the collective bargaining agent; in the absence of such an agent, by representatives of the employees in the affected unit.

9. The plan will not serve as a subsidy of seasonal employment during the off season, nor as a subsidy of temporary part-time or intermittent employment.

10. The employer agrees to furnish reports relating to the proper conduct of the plan and agrees to allow the Director or his/her authorized representatives access to all records necessary to verify the plan prior to approval and, after approval, to monitor and evaluate application of the plan.

In addition to the matters specified above, the Director shall take into account any other factors which may be pertinent to proper implementation of the plan.

C. *Approval or Rejection of the Plan*

The Director shall approve or reject a plan in writing within ___ days of its receipt. The reasons for rejection shall be final and non-appealable, but the employer shall be allowed to submit another plan for approval not earlier than ___ days from the date of the earlier rejection.

D. *Effective Date and Duration of Plan*

A plan shall be effective on the date specified in the plan or on a date mutually agreed upon by the employer and the Director. It shall expire at the end of the 12th full calendar month after its effective date or on the date specified in the plan if such date is earlier, provided that the plan is not previously revoked by the Director. If a plan is revoked by the Director, it shall terminate on the date specified in the Director's written order of revocation.

E. *Revocation of Approval*

The Director may revoke approval of a plan for good cause. The revocation order shall be in writing and shall specify the date the revocation is effective and the reasons therefor.

Good cause shall include, but not be limited to, failure to comply with the assurances given in the plan, unreasonable revision of productivity standards for the affected unit, conduct or occurrences tending to defeat the intent and effective operation of the plan, and violation of any criteria on which approval of the plan was based.

Such action may be taken at any time by the Director on his/her own motion, on the motion of any of the affected unit's employees or on the motion of the appropriate collective bargaining agent(s); provided that the Director shall review the operation of each qualified employer plan at least once during the 12-month period the plan is in effect to assure its compliance with the requirements of these provisions.

F. *Modification of an Approved Plan*

An operational approved STC plan may be modified by the employer with the acquiescence of employee representatives if the modification is not substantial and in conformity with the plan approved by the Director, but the modifications must be reported promptly to the Director. If the hours of work are increased or decreased substantially beyond the level in the original plan, or any other conditions are changed substantially, the Director shall approve or disapprove such modifications, without changing the expiration date of the original plan. If the substantial modifications do not meet the requirements for approval, the Director shall disallow that portion of the plan in writing as specified in section E.

G. *Eligibility for Short-Time Compensation*

1. An individual is eligible to receive STC benefits with respect to any week only if, in addition to monetary entitlement, the Director finds that:

(a) During the week, the individual is employed as a member of an affected unit under an approved short-time compensation plan which was approved prior to that week, and the plan is in effect with respect to the week for which STC is claimed.

(b) The individual is able to work and is available for the normal work week with the short-time employer.

(c) Notwithstanding any other provisions of this Act to the contrary, an individual is deemed unemployed in any week for which remuneration is payable to him/her as an employee in an affected unit for 90 percent or less than his/her normal weekly hours of work as specified under the approved short-time compensation plan in effect for the week.

(d) Notwithstanding any other provisions of this Act to the contrary, an individual shall not be denied STC benefits for any week by reason of the application of provisions relating to availability for work and active search for work with an employer other than the short-time employer.

H. *Benefits*

1. The short-time weekly benefit amount shall be the product of the regular weekly unemployment compensation amount multiplied by the percentage of reduction of at least 10 percent in the individual's usual weekly hours of work.

2. An individual may be eligible for STC benefits or unemployment compensation, as appropriate, except that no individual shall be eligible for combined benefits in any benefit year in an amount more than the maximum entitlement established for unemployment compensation, nor shall an individual be paid STC benefits for more than 26 weeks (whether or not consecutive) in any benefit year pursuant to a short-time plan.

3. The STC benefits paid an individual shall be deducted from the maximum entitlement amount established for that individual's benefit year.

4. Claims for STC benefits shall be filed in the same manner as claims for unemployment compensation or as prescribed in regulations by the Director.

5. Provisions applicable to unemployment compensation claimants shall apply to STC claimants to the extent that they are not inconsistent with STC provisions. An individual who files an initial claim for STC benefits shall be provided, if eligible therefor, a monetary determination of entitlement to STC benefits and shall serve a waiting week.

6. (a) If an individual works in the same week for an employer other than the short-time employer and his or her combined hours of work for both employers are equal to or greater than the usual hours of work with the short-time employer, he or she shall not be entitled to benefits under these short-time provisions or the unemployment compensation provisions.

(b) If an individual works in the same week for both the short-time employer and another employer and his or her combined hours of work for both employers are equal to or less than 90 percent of the usual hours of work for the short-time employer, the benefit amount payable for that week shall be the weekly unemployment compensation amount reduced by the same percentage that the combined hours are of the usual hours of work. A week for which benefits are paid under this provision shall count as a week of short-time compensation.

(c) If an individual did not work during any portion of the work week, other than the reduced portion covered by the short-time plan, with the approval of the employer, he or she shall not be disqualified for such absence or deemed ineligible for STC benefits for that reason alone.

7. An individual who performs no services during a week for the short-time employer and is otherwise eligible, shall be paid the full weekly unemployment compensation amount. Such a week shall not be counted as a week with respect to which STC benefits were received.

8. An individual who does not work for the short-time employer during a week but works for another employer and is otherwise eligible, shall be paid benefits for that week under the partial unemployment compensation provisions of the State law. Such a week shall not be counted as a week with respect to which STC benefits were received.

I. *Charging Shared Work Benefits*

STC benefits shall be charged to employers' experience rating accounts in the same manner as unemployment compensation is charged under the State law. Employers liable for payments in lieu of contributions shall have STC benefits attributed to service in their employ in the same manner as unemployment compensation is attributed.

J. *Extended Benefits*

An individual who has received all of the STC benefits or combined unemployment compensation and STC benefits available in a benefit year shall be considered an exhaustee for purposes of extended benefits, as provided under the provisions of section ___, and, if otherwise eligible under those provisions, shall be eligible to receive extended benefits.

Further information, including commentary on the model legislative language, is available from the U.S. Department of Labor, Employment and Training Administration, Washington, D.C. 20213.

Index

About the Editors and Contributors

EDITORS

RAMELLE MACOY is president of Collective Bargaining Associates, consultants to unions. He served as director of the Association of Pennsylvania State College and University Faculties (1976–1981) and has held various union leadership positions. These include three years as the national director of organization for the Allied Industrial Workers' Union and two years as the union's national director of education. He has also acted as regional director for the American Federation of State, County, and Municipal Employees, Rocky Mountain Region, and associate regional director for the International Ladies' Garment Workers' Union, Southeast Region. In addition to his union activities, MaCoy has had an extensive career in journalism, including the editorship of the Shippensburg (Pa.) *News-Chronicle* and eight years (1948–1956) as foreign correspondent for *Time* and *Life* magazines in South and Central America, Eastern Europe, Japan, and Korea. For two of those years he acted as South American bureau chief in Buenos Aires, Argentina. MaCoy received his B.S. degree from the University of Illinois and has done graduate work at Mexico City College.

MARTIN J. MORAND is a professor in the graduate department of industrial and labor relations at Indiana University of Pennsylvania and also serves as director of the university's Pennsylvania Center for the Study of Labor Relations. Since 1977 he has been writing and researching in the area of short-time compensation. Rep. Patricia Schroeder (D., CO) has credited Morand's work as a major contribution to the passage of her federal short-time compensation legislation. More recently he has provided information and testimony in support of STC legislation in Pennsylvania and other states and continues to consult with legislators and administrators on work-sharing unemployment insurance in California, Arizona, Oregon, and Canada. Morand received his B.S. degree from the New York State School of Industrial and Labor Relations at Cornell University. Upon graduation, Morand held various union leadership positions. From 1948 to 1969 he served as organizer, business agent, education director, and Southeast regional director for the International Ladies' Garment Workers' Union. During the next few years, he was education director of the American Federation of State, County and Municipal Employees; consultant at the George Meany Center for Labor Studies; and director of the Association of Pennsylvania State College and University Faculties.

CONTRIBUTORS

RUTH GERBER BLUMROSEN is associate professor at the Graduate School of Management, Rutgers University, and a labor arbitrator and consultant specializing in the area of equal employment opportunity. She received her J.D. degree from the University of Michigan Law School and is a member of the Michigan bar, the U.S. Supreme Court bar, and the Third Circuit Court of Appeals bar. Formerly an assistant dean at the Howard University Law School, Blumrosen has served as consultant to the Department of Health and Human Services, the Equal Opportunity Commission (EEOC), and to former EEOC Chairperson Eleanor Holmes Norton. As an EEOC consultant, she assisted in the examination of wage discrimination problems and the impact of layoffs on affirmative action programs. Blumrosen's experience also includes extensive speaking and writing in the areas of equal employment opportunity and work sharing.

DONNA HUNTER is manager of program control for the Employment Division, Department of Human Resources, State of Oregon. Over the past eight years she has been active in presenting unemployment insurance workshops for both employers and union business agents. She has also acted as a consultant to the U.S. Department of Labor, designing and presenting national training packages on new federal programs and projects. Hunter is a member of Oregon's legislative team and worked closely with the parties who succeeded in passing short-time compensation legislation, which established the Workshare program. Her department was responsible for the implementation of Workshare and is currently evaluating the program to determine whether future legislative changes are needed.

LINDA A. ITTNER works for the Select Committee on Children, Youth, and Families and staffs the Economic Security Task Force, chaired by Rep. Patricia Schroeder (D., CO). From 1977 to 1983 she was a staff member for the Civil Service Subcommittee, specializing in alternative work schedules legislation and civil service issues, particularly employee participation and productivity enhancement in the federal government. Previously, she worked in the office of Rep. Schroeder on pension reforms to benefit spouses of federal workers covered by civil service, foreign service, military, and CIA retirement systems.

JOHN LAMMERS is currently a postdoctoral fellow in organizations research and training in the sociology department at Stanford University. Together with co-author Timothy Lockwood, he served as a researcher with the California evaluation of the Shared Work Unemployment Insurance program (SWUI) and is also a founding member of the Davis Labor Process Group, an organization of sociologists conducting research on changes in the contemporary workplace and in employment policy. Lammers' dissertation for his Ph.D., which he received from the University of California at Davis, is an extended analysis of the sources, impacts, and efficacy of California's SWUI program. His areas of specialization include political sociology, formal organizations, and policy research and evaluation.

TIMOTHY LOCKWOOD is a Ph.D. candidate at the University of California at Davis. With co-author John Lammers, he was a researcher with the California evaluation of the Shared Work Unemployment Insurance program and is a founding member of the Davis Labor Process group. Lockwood's research interests are in modern leisure and its relationship to ongoing changes in labor markets. He is also studying changes in blue-collar culture that result from plant closings and subsequent worker reemployment. His specializations are political sociology, the sociology of labor markets, occupational communities, and urban sociology.

HARRY MEISEL is head of the Federal Employment Institution in the State of Baden-Wurttemberg, West Germany, which directs the unemployment insurance scheme, employment training and retraining programs, and vocational guidance and placement services. He was elected to his present position by a tripartite body (management, labor, and state and local government), after having served in different fields of responsibility in his country and abroad. He holds an LL.D. degree from Innsbruck University in Austria and has authored publications on labor-market problems, vocational guidance, and training of the handicapped.

NOAH M. MELTZ is professor of economics and director of the Centre for Industrial Relations, University of Toronto. He received his Ph.D. from Princeton University. His research interests include labor-market analysis and manpower and industrial relations. Meltz is a past president of the Canadian Industrial Relations Association, and his advisory appointments have included the Canada Department of Employment and Immigration, the Ontario Manpower Commission, and the Israel Central

Bureau of Statistics. His recently published books are: *Lagging Productivity Growth: Causes and Remedies* (1980, with Shlomo Maital); *Sharing the Work: An Examination of the Issues in Worksharing and Jobsharing* (1981, with Frank Reid and Gerald Swartz); *Economic Analysis of Labour Shortages: The Case of Tool and Die Makers in Ontario* (1982); and *Personnel Management in Canada* (1983, with Thomas Stone).

MARTIN NEMIROW is a social science advisor for the U.S. Department of Labor, Washington, D.C. He served as a senior policy analyst in the Office of the Secretary for most of 1967 to 1982. He has specialized in measures to deal with unemployment and related social policies. Nemirow's paper on blue-collar and lower-middle-income workers was used by Assistant Secretary of Labor Jerome M. Rosow as the basis for a report to President Nixon that became a framework for certain policy initiatives in that period. He has been responsible for the initiation of federal policy development in such areas as short-time compensation and protection of worker privacy. He has served as the chief analyst for the Labor Department on higher education policy, national health insurance, and health insurance for the unemployed. In 1973–74, Nemirow was appointed a Congressional Fellow, serving as legislative aide to Senator Hubert Humphrey, and then to Rep. John B. Anderson. He is a former senior editor for Prentice-Hall and a graduate of Harvard University.

FRANK REID is an associate professor in economics at the Centre for Industrial Relations, University of Toronto. He obtained his M.S. in economics at the London School of Economics and his Ph.D. at Queen's University. His publications include *Wage and Price Behaviour in Canadian Manufacturing* (1979, with T. A. Wilson), *Sharing the Work* (1981, with Noah M. Meltz and Gerald Swartz), and *Sex Discrimination in the Canadian Labour Market* (1983, with M. Gunderson). His research interests cover a wide range of labor-market topics, and his current project is an analysis of the causes and consequences of strike activity in Canada.

ROBERT ST. LOUIS, an associate professor in the College of Business Administration at Arizona State University, received his Ph.D. in economics from Purdue University. Before joining the faculty at ASU, he was chief of research for the Arizona Unemployment Insurance Administration and chief economic-forecast analyst for the Arizona Executive Budget Office. While working for the Unemployment Insurance Administration, one of his major responsibilities was to assess the feasibility and desirability of introducing a shared-work unemployment compensation program in Arizona. He worked extensively with legislators, employers, employees, and public representatives in order to determine what form of shared-work unemployment compensation program would be most beneficial to the people of Arizona. The results of that research provided the basis for the formulation of Senate Bill 1005 (Arizona's original shared-work bill).